KV-684-939

SMP interact

T1

Teacher's guide to Book T1

PUBLISHED BY THE PRESS SYNDICATE OF THE UNIVERSITY OF CAMBRIDGE
The Pitt Building, Trumpington Street, Cambridge, United Kingdom

CAMBRIDGE UNIVERSITY PRESS
The Edinburgh Building, Cambridge CB2 2RU, UK
40 West 20th Street, New York, NY 10011–4211, USA
10 Stamford Road, Oakleigh, VIC 3166, Australia
Ruiz de Alarcón 13, 28014 Madrid, Spain
Dock House, The Waterfront, Cape Town 8001, South Africa

http://www.cambridge.org

© The School Mathematics Project 2001
First published 2001
Reprinted 2001

Printed in the United Kingdom at the University Press, Cambridge

Typeface Minion   *System* QuarkXPress®

A catalogue record for this book is available from the British Library

ISBN 0 521 79861 2 paperback

Typesetting and technical illustrations by The School Mathematics Project
Cover image © Tony Stone Images/Darryl Torckler
Cover design by Angela Ashton

NOTICE TO TEACHERS
It is illegal to reproduce any part of this work in material form
(including photocopying and electronic storage) except under the
following circumstances:
(i)   where you are abiding by a licence granted to your school or
      institution by the Copyright Licensing Agency;
(ii) where no such licence exists, or where your wish to exceed the
      terms of a licence, and you have gained the written permission of
      Cambridge University Press;
(iii) where you are allowed to reproduce without permission under the
      provisions of Chapter 3 of the Copyright, Designs and Patents Act 1988.

Contents

Introduction		5
General guidance on teaching approaches		7
1	Time	9
2	Action and result puzzles	12
3	Chance	15
4	Symmetry	18
5	Decimals 1	22
	Review 1	27
	Mixed questions 1	27
6	Number grids	28
7	Stars and angles	36
8	Lines at right angles	38
9	Comparisons	40
10	Ice cream	45
11	Parallel lines	46
12	Anamorphs	48
	Review 2	49
	Mixed questions 2	49
13	Practical problems	50
14	Angle dominoes	53
15	Quadrilaterals	54
16	Is it an add?	57
17	Desk tidy	59
18	Frequency	60
19	Amazing but true!	65
	Review 3	66
	Mixed questions 3	66
20	Photo display	67
21	Fractions 1	68
22	Enlargement	70
23	Calculate in order	72
24	Graphs and charts	76
25	Fractions 2	81
26	Negative numbers	85
	Review 4	88
	Mixed questions 4	88

The following people contributed to the writing of the SMP Interact key stage 3 materials.

Ben Alldred	Ian Edney	John Ling	Susan Shilton
Juliette Baldwin	Steve Feller	Carole Martin	Caroline Starkey
Simon Baxter	Rose Flower	Peter Moody	Liz Stewart
Gill Beeney	John Gardiner	Lorna Mulhern	Pam Turner
Roger Beeney	Bob Hartman	Mary Pardoe	Biff Vernon
Roger Bentote	Spencer Instone	Peter Ransom	Jo Waddingham
Sue Briggs	Liz Jackson	Paul Scruton	Nigel Webb
David Cassell	Pamela Leon	Richard Sharpe	Heather West

Others, too numerous to mention individually, gave valuable advice, particularly by commenting on and trialling draft materials.

Editorial team:	David Cassell	Project Administrator:	Ann White
	Spencer Instone	Design:	Melanie Bull
	John Ling		Tiffany Passmore
	Mary Pardoe		Martin Smith
	Paul Scruton	Project support:	Carol Cole
	Susan Shilton		Pam Keetch
			Nicky Lake
			Jane Seaton
			Cathy Syred

Special thanks go to Colin Goldsmith.

Introduction

What is distinctive about *SMP Interact*?

SMP Interact sets out to help teachers use a variety of teaching approaches in order to stimulate pupils and foster their understanding and enjoyment of mathematics.

A central place is given to discussion and other interactive work. Through discussion with the whole class you can find out about pupils' prior understanding when beginning a topic, can check on their progress and can draw ideas together as work comes to an end. Working interactively on some topics in small groups gives pupils, including the less confident, a chance to clarify and justify their own ideas and to build on, or raise objections to, suggestions put forward by others.

Questions that promote effective discussion and activities well suited to group work occur throughout the material.

SMP Interact has benefited from extensive and successful trialling in a variety of schools. The practical suggestions contained in the teacher's guides are based on teachers' experiences, often expressed in their own words.

Who are Books T1 to 3 for?

Books T1 to 3 follow on from Books 1 and N and cover national curriculum levels up to 5.

How are the pupils' books intended to be used?

The pupils' books are a resource which can and should be used flexibly. They are not for pupils to work through individually at their own pace. Many of the activities are designed for class or group discussion.

Activities intended to be led by the teacher are shown by a solid strip at the edge of the pupil's page, and a corresponding strip in the margin of the teacher's guide, where they are fully described.

A broken strip at the edge of the page shows an activity or question in the pupil's book that is likely to need teacher intervention and support.

Where the writers have a particular way of working in mind, this is stated (for example, 'for two or more people').

Where there is no indication otherwise, the material is suitable for pupils working on their own.

Starred questions (for example, *C7) are more challenging.

What use is made of software?

Points at which software (on a computer or a graphic calculator) can be used to provide effective support for the work are indicated by these symbols, referring to a spreadsheet, graph plotter or dynamic geometry package respectively. Other suggestions for software support can be found on the SMP's website: www.smpmaths.org.uk

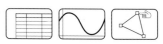

How is the attainment of pupils assessed?

The interactive class sessions provide much feedback to the teacher about pupils' levels of understanding.

Each unit of work begins with a statement of the key learning objectives and finishes with questions for self-assessment ('What progress have you made?') The latter can be incorporated into a running record of progress.

Revision questions are included in periodic reviews in the pupil's book.

A pack of assessment materials for Books T1, S1 and C1 contains photocopiable masters providing a short assessment for most of the units. Enclosed with the pack is a CD-ROM holding the assessment materials in question bank form so you can compile and edit tests on screen to meet your school's needs. Details of the pack are on the SMP's website.

What will pupils do for homework?

The practice booklets may be used for homework.

Often a homework can consist of preparatory or follow-up work to an activity in the main pupil's book.

Answers to questions on resource sheets

For reasons of economy, where pupils have to write their responses on a resource sheet the answers are not always shown in this guide. For convenience in marking you could put the correct responses on a spare copy of each sheet and add it to a file for future use.

General guidance on teaching approaches

Getting everyone involved

When you are introducing a new idea or extending an already familiar topic, it is important to get as many pupils as possible actively engaged.

Posing key questions
A powerful technique for achieving this is to pose one or two key questions, perhaps in the form of a novel problem to be solved. Ask pupils, working in pairs or small groups, to think about the question and to try to come up with an answer.

When everyone has had time to work seriously at the problem (have a further question ready for the faster ones), you can then ask for answers, without at this stage revealing whether they are right or wrong (so that pupils have to keep thinking!). You could ask pupils to comment on each other's answers.

Open tasks
Open tasks and questions are often good for getting pupils to think, or to think more deeply. For example, 'Working in groups of three or four, make up three questions which can be solved using the technique we have just been learning. Try to make your questions as varied as possible.'

Questioning skills

Questioning with the whole class
If your questions to the class are always closed, and you reward the first correct response you get, then you have no way of telling whether other pupils knew the correct answer or whether they had thought about the question at all. It is better to try to get as many pupils as possible to engage with the question, so do not at first say whether an answer is right or wrong. You could ask a pupil how they got their answer, or you could ask a second pupil how they think the first one got their answer.

Working in groups

Types of group work
Group work may be small scale or large scale. In small scale group work, pupils are asked to work in pairs or small groups for a short while, perhaps to come up with a solution to a novel type of problem before their suggestions are compared. In large scale group work, pupils carry out in groups a substantial task such as planning a statistical inquiry or designing a poster to get over the essential idea of the topic they have just been studying.

Organising the groups
Group size is important. Groups of more than four or five can lead to some pupils making little or no contribution.

For some activities, you may want pupils to work unassisted. But for many, your own contribution will be vital. Then it is generally more effective if, once you are sure that every group has got started, you work intensively with each group in turn.

After the group work One way to help pupils feel that they have worked effectively is to get them to report their findings to the whole class. This may be done in a number of different ways. One pupil from each group could report back. Or you could question each group in turn. Or each group could make a poster showing their results.

Managing discussion

Discussion, whether in a whole-class or group setting, has a vital role to play in developing pupils' understanding. It is most fruitful in an atmosphere where pupils know their contributions are valued and are not always judged in terms of immediate correctness. It needs careful management for it to be effective and teachers are often worried that it will get out of hand. Here are a few common worries, and ways of dealing with them.

What if ... '... the group is not used to discussion?'

- Allow time for pupils to work first on the problem individually or in small groups, then they will all have ideas to contribute.

'... everyone tries to talk at once?'

- Set clear rules. For example, pupils raise their hands and you write their name on the board before they can speak.

'... a few pupils dominate whole-class discussion?'

- Precede any class discussion with small-group discussion and nominate the pupils who will feed back to the class.

'... one pupil reaches the end point of a discussion immediately?'

- Tell them that the rest of the group need to be convinced and ask the pupil to convince the rest of the group.

- Accept the suggestion and ask the rest of the group to comment on it.

1 Time

T	p 4 **A** Happiness graphs	Introductory discussion and using a time line
T	p 4 **B** Time planner	Further use of a time line
T	p 5 **C** At the same time	Ways of writing equivalent times
	p 5 **D** Time lines	Calculating time intervals, given a time line
T	p 6 **E** How long?	Calculating time intervals over one hour

Essential	Optional
Sheet 97	Sheet 98 or 99
Scissors	OHP transparency of sheet 98 or 99
Practice booklet pages 3 to 5	

△ **Happiness graphs** (p 4)

> Optional: Sheet 98 or 99, and a transparency of the sheet used

◊ Discuss the 'happiness graph'. Happiness is 'measured' on a 0 to 10 scale.

◊ Pupils can then draw their own happiness graphs on squared paper or on specially ruled time graph paper (sheet 98 or 99).

'Very good. I did my own happiness graph on the board for the day I taught the class.'

Sheet 98

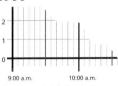

Sheet 99

Pupils should label the axes as appropriate.

◊ Before pupils draw their graphs it may be necessary to establish the times of daily events (e.g. lesson changes, breaks). A transparency of the time graph paper is useful here.

ℬ **Time planner** (p 4)

> Optional: Sheet 98 or 99, and a transparency of the sheet used

◊ Discuss the diagram at the bottom of page 4, posing questions such as 'When does assembly end?', 'How long is break?'

◊ Discuss how you could devise a time plan for a whole week by producing a set of bars, one for each day. Pupils can draw diagrams to show their own timetables (including the weekend if they like). A shorter activity is to produce a diagram for the current day only.

If the whole group has the same timetable, each pupil could do one particular school day and then the days could be collected together to make weekly timetables.

Alternatively, each pupil could show a typical Saturday or Sunday.

As in section A, they can consider the day from 9 a.m. to 4 p.m. (sheet 98) or from 9 a.m. to 11 p.m. (sheet 99).

C **At the same time** (p 5)

> Sheet 97, scissors

◊ If you haven't already done so in sections A or B, discuss how to convert times like a quarter to seven in the morning to 6:45 a.m., and other equivalent forms.

Pupils then order the times on sheet 97. They are a set of cards that pupils can cut out and put in order. This requires quite a lot of desk space, and it may be useful for pupils to work in pairs.

The correct order is: B, L, Q, A, M, R, O, D, N, K, G, I, E, C, P, J, F, H.

◊ A simple game can be played in groups of three.
- Shuffle the cards and deal six each.
- Each player plays a card.
- The latest time (or earliest, or middle, as agreed) wins the trick. The winner of the trick goes first in the next round.

D **Time lines** (p 5)

D9 You may need to remind pupils of the work they did on time planners in section B, where time intervals were presented as here.

E **How long?** (p 6)

◊ You might find it more relevant for the pupils if you discuss how to work out the lengths of some of their lessons, providing that this brings up the problem of working out an interval over two different hours.

TV listings from newspapers can easily be used as the basis for further real-life examples.

E1, 2 Pupils may find it useful to sketch a time line similar to that in the introductory example.

D Time lines (p 5)

D1 3:20 p.m.

D2 3:40 p.m.

D3 20 minutes

D4 4:10 p.m.

D5 30 minutes

D6 4:50 p.m.

D7 5:30 p.m.

D8 (a) 30 minutes (b) 1 hour 10 minutes

D9 1:20 p.m.

D10 30 minutes

D11 1 hour

D12 1 hour 30 minutes

D13 3:45 p.m.

E How long? (p 6)

E1 (a) 1 hour 40 minutes
   (b) 2 hours 30 minutes
   (c) 2 hours 50 minutes
   (d) 1 hour 50 minutes

E2 (a) 1 hour 45 minutes
   (b) 1 hour 40 minutes
   (c) 1 hour 45 minutes
   (d) 1 hour 30 minutes

E3 1 hour 40 minutes

E4 35 minutes

E5 15 minutes

E6 35 minutes

E7 20 minutes

E8 1:45 p.m.

E9 3:25 p.m.

E10 35 minutes

E11 (a) 1 hour 15 minutes
   (b) 1 hour 40 minutes

E12 10 minutes

E13 10:35

What progress have you made? (p 7)

1 (a) 45 minutes   (b) 1 hour 20 minutes
   (c) 1 hour 20 minutes
   (d) 1 hour 45 minutes

Practice booklet

Section C (p 3)

1 A and G, B and J, C and H,
   D and K, E and L, F and I

2 (a) 6:30 a.m. (b) 10:15 p.m. (c) 8:45 p.m.
   (d) 7:40 a.m. (e) 9:35 p.m. (f) 9:50 a.m.
   (g) 2:55 p.m. (h) 5:05 a.m.

Sections D and E (p 4)

1 (a) 15 minutes   (b) 25 minutes
   (c) 55 minutes   (d) 20 minutes
   (e) 1 hour 25 minutes
   (f) 1 hour 30 minutes
   (g) 1 hour 30 minutes
   (h) 6 hours 20 minutes
   (i) 2 hours 35 minutes

2 (a) 20 minutes   (b) 1 hour 15 minutes
   (c) 20 minutes   (d) 50 minutes
   (e) 2 hours 45 minutes

3 (a) 45 minutes   (b) 1 hour 20 minutes
   (c) 1 hour 20 minutes
   (d) 1 hour 45 minutes
   (e) 2 hours 45 minutes
   (f) 1 hour 20 minutes

4 (a) 45 minutes   (b) 10 minutes
   (c) 1 hour 30 minutes

5 (a) 30 minutes   (b) 25 minutes
   (c) 20 minutes   (d) 45 minutes

Action and result puzzles (p 8)

In each puzzle, the action cards show operations to be performed on a starting number and the result cards show the results. Pupils match up the results with the actions.

The puzzles provide number practice and an opportunity to apply some logical thinking. They may reveal misconceptions about number.

Essential	**Optional**
Puzzles on sheets 102 to 105	Sheet 110 (blank cards)
Scissors	Puzzles on sheets 106 and 107 (harder)
	OHP transparencies of some sheets, cut into puzzle cards
Practice booklet pages 6 and 7	

T

The puzzles are listed below, roughly in order of difficulty. Some weaker pupils may have difficulty with the puzzles on sheet 105.

'The letter games were really enjoyed by the more able in the group and provoked good discussion and sharing methods. They homed in on the correct method together.'

Sheet 102 6 puzzle (+ and −, whole numbers ≤ 10)
31 puzzle (+ and −, whole numbers ≤ 50)
4 puzzle (× and ÷ by 1, 2, 4 and 5)

Sheet 103 5 puzzle (+, −, × and ÷, whole numbers ≤ 20)
16 puzzle (+, −, × and ÷, whole numbers ≤ 32)
270 puzzle (+ and −, multiples of 10)

Sheet 104 44 puzzle (+ and −, whole numbers ≤ 100)
60 puzzle (+, −, × and ÷, simple two- and three-digit numbers)
5.5 puzzle (+ and −, decimals with 0.5 only)

Sheet 105 7.5 puzzle (+, − and × 2, with simple decimals)
p puzzle ('logic' puzzle, + and −, one- and two-digit numbers)
m puzzle ('logic' puzzle, +, −, × and ÷, two-digit numbers)

Pupils who do well with these can do the following as extension.

Sheet 106 327 puzzle (+ and −, three-digit numbers)
6.5 puzzle (+ and −, includes simple fractions and decimals)
3679 puzzle (+ and −, four-digit numbers)

Sheet 107 36 puzzle (+, −, × and ÷, includes simple decimals and negative numbers)
q puzzle ('logic' puzzle, + and −, two-digit numbers)
h puzzle ('logic' puzzle, +, −, × and ÷, two-digit numbers)

'I copied the cards on to pieces of acetate which could be moved about on the OHP. Pupils went to the OHP to show how the cards matched up.'

◊ This has worked well with pupils sitting in pairs on tables of four. When each pair has matched the cards, all four pupils discuss what they have done. An aim is to encourage mental number work. However, pupils may want to do some calculations and demonstrate things to their group using pencil and paper. It is not intended that a calculator should be used.

◊ Puzzles that pupils find easy can be done without cutting out the cards: they simply key each action card to its result card by marking both with the same letter. However, something may be learnt from moving cards around to try ideas out before reaching a final pairing, and some puzzles are almost impossible unless they are done this way.

◊ Solutions can be recorded by
- keying cards to one another with letters as described above
- sticking pairs of cards on sheets or in exercise books
- writing appropriate statements, such as 8 − 3 = 5

◊ After pupils have solved some puzzles, they can make up some of their own (using the blank cards) to try on a partner. This may tell you something about the limits of the mathematics they feel confident with. Some should be able to make up puzzles of the *p* and *m* type.

Sheet 102

6 puzzle

Action	Result
− 2	4
+ 3	9
− 4	2
+ 4	10
+ 2	8
− 3	3

31 puzzle

Action	Result
+ 10	41
− 10	21
+ 9	40
− 9	22
− 19	12
+ 19	50
− 7	24
+ 15	46

4 puzzle

Action	Result
× 2	8
÷ 2	2
× 4	16
÷ 4	1
÷ 1	4
× 5	20

Sheet 103

5 puzzle

Action	Result		
+ 10	15	× 2	10
× 4	20	× 3	15
− 3	2	+ 5	10
+ 7	12	÷ 1	5

16 puzzle

Action	Result
÷ 4	4
− 4	12
+ 4	20
÷ 8	2
− 8	8
× 2	32
÷ 2	8
+ 5	21

270 puzzle

Action	Result
+ 100	370
− 80	190
+ 300	570
− 60	210
+ 30	300
− 150	120
+ 240	510
− 100	170

Sheet 104

44 puzzle

Action	Result
− 43	1
+ 43	87
− 29	15
+ 29	73
+ 50	94
− 9	35
+ 56	100
− 14	30

60 puzzle

Action	Result
÷ 10	6
− 10	50
− 12	48
÷ 5	12
× 3	180
× 10	600
+ 100	160
÷ 3	20

5.5 puzzle

Action	Result		
+ 0.5	6	− 1.5	4
− 0.5	5	+ 9	14.5
+ 2.5	8	+ 3.5	9
− 2.5	3	− 3.5	2

Sheet 105

7.5 puzzle

Action	Result
− 0.3	7.2
× 2	15
− 2.5	5
+ 0.5	8
+ 1.5	9
− 0.5	7
+ 10	17.5
+ 0.3	7.8

p puzzle: $p = 5$

m puzzle: $m = 12$

Sheet 106

327 puzzle

Action	Result
− 30	297
+ 40	367
− 110	217
+ 700	1027
+ 390	717
− 89	238
+ 651	978
− 207	120

6.5 puzzle

Action	Result
+ 2.5	9
− 3.5	3
+ 9	15.5
+ 2.25	8.75
− 0.5	6
+ 4.75	11.25
$-1\frac{1}{2}$	5
$-\frac{3}{4}$	5.75

3679 puzzle

Action	Result
+ 30	3709
− 2030	1649
+ 45	3724
− 95	3584
− 2300	1379
+ 4205	7884
+ 999	4678
− 680	2999

Sheet 107

36 puzzle

Action	Result	Action	Result
÷ 9	4	÷ 8	4.5
− 40	⁻4	× 1.5	54
× 3	108	+ ⁻50	⁻14
+ 27	63	÷ 24	1.5

q puzzle: $q = 27$

h puzzle: $h = 9$

Practice booklet (p 6)

1 15 puzzle

Action	Result
+ 3	18
+ 7	22
− 7	8
− 3	12
− 10	5
+ 5	20
− 5	10
+ 10	25

2 (a) 6   (b) 24

3 (a) add 2   (b) add 6   (c) subtract 11

4 20 puzzle

Action	Result
× 2	40
+ 11	31
− 7	13
− 13	7
÷ 2	10
÷ 5	4
× 5	100
+ 30	50

5 (a) 11   (b) 30   (c) 5   (d) 120

6 (a) subtract 5   (b) add 13

(c) add 60 *or* multiply by 4

7 add 40, *and* multiply by 3

Chance

This unit introduces probability through games of chance. Probability is expressed as a fraction in situations with equally likely outcomes.

p 9	**A** Chance or skill?	Deciding whether games are skill or chance
p 10	**B** Fair or unfair?	Deciding whether a game of chance is fair
p 11	**C** Probability	The probability scale from 0 to 1
p 12	**D** Spinners	Writing a probability as a fraction
		Finding the probability of an event not happening

Essential	**Optional**
Dice, counters of different colours	OHP transparencies of sheets 111 to 115
Sheets 111 to 115	
Practice booklet pages 8 and 9	

Ⓐ **Chance or skill?** (p 9)

> Dice and counters
> Sheets 111 to 113 (game boards)

◊ Before discussing and playing the games, you could get pupils talking about chance, for example, the National Lottery. People often have peculiar ideas about chance. For example, would they write on a National Lottery ticket the same combination as the one that won last week? If not, why not?

  You could ask pupils to think about games that they know and to discuss the elements of chance and skill in them.

◊ It may be useful to copy the games on to card and laminate them or cover them with clear adhesive film. The games can then be used more than once.

◊ Before playing each game, ask pupils to try to decide from its rules whether it is a game of pure chance, a game of skill, or a mixture.

  Some games of skill give an advantage to the first player. Who goes first is usually decided by a process of chance.

'Some pupils thought that using a dice meant it was all chance.'

◊ You could split the class into pairs or small groups, with each group playing one of the games and reporting on it.

◊ 'Fours' is a game of skill. 'Line of three' is a mixture of chance and skill. 'Jumping the line' appears to involve skill, because you have to decide which counters to move and it looks as if you can get 'nearer' to winning. But it is a game of pure chance. At any stage there is only one number which will enable the player to win. If any other number comes up, whatever the player does leaves the opponent in essentially the same position.

B **Fair or unfair?** (p 10)

Pupils have to decide whether a game of chance is fair or unfair.

> Dice, counters, sheets 114 and 115
> Optional: Transparencies of sheets 114 and 115

◊ You can start by playing the game several times as a class, with a track on the board. You may need to go over the meaning of 'odd' and 'even'.

◊ When pupils play the game themselves, ask them to record the results and then pool the class's results.

◊ Let pupils consider each other's ways of making the game fairer (if they can think of any!). Do they agree that they would be fairer?

'I split the class into groups and gave rat numbers to each group. There was intense rivalry!'

◊ For the first rat race, counters ('rats') are lined up at the start. The teacher rolls a dice. The score tells which rat moves forward one square. Everyone chooses a rat they think will win. For the second rat race, two dice are rolled and the total is used.

In the first race, some pupils may believe that 6 is 'harder' to get than other numbers. If so, you could discuss this.

In the second race you could ask for suggestions for making it fairer, still using two dice. (For example, the track could be shortened for the 'end' numbers – even so, Rat 1 is never going to win!)

C **Probability** (p 11)

The probability scale from 0 to 1 is introduced.

◊ Construct a probability 'washing line' by pinning the ends of a long piece of string to the board. Mark 0 and 1 at the ends of the line. Pupils can then be asked to hang cards for different events on the line with paper clips. Explain first the meanings of the two endpoints of the scale. Something with probability 0 is often described as 'impossible'. However, there are different ways of being impossible and some of them have nothing to do with probability (for example, it is impossible for a triangle to have four sides). So it is better to say 'never happens'. Something with probability 1 always happens, or is certain to happen.

◊ Go through the events listed in the pupil's book and discuss where they go on the scale. The coin example leads to the other especially important point on the scale, $\frac{1}{2}$. Associate this with 'equally likely to happen or not happen', with fairness, 'even chances', etc.

◊ Keep the approach informal. The important thing is to locate a point on the right side of $\frac{1}{2}$, or close to one of the ends when appropriate (for example, in the case of the National Lottery!).

Ⅾ **Spinners** (p 12)

Pupils write probabilities as fractions.

◊ A spinner is very useful in connection with probability. It shows fractions in a familiar way. This is a good opportunity to revise and reinforce basic fraction work. Emphasise that the parts (sectors) into which the spinner is divided have to be equal.

D3 If pupils give $\frac{1}{3}$ for (d), they have ignored the inequality of the parts.

Make your own spinner (p 13)

This is a good opportunity to do some further tallying and bar charts, as well as accurate drawing. The shape of the bar chart will show how fair a spinner is.

Ⅽ **Probability** (p 11)

C1

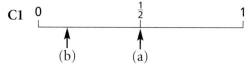

C2 (a) A boy

(b)

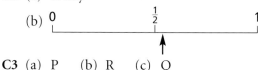

C3 (a) P (b) R (c) Q

Ⅾ **Spinners** (p 12)

D1 (a) $\frac{1}{2}$ (b) $\frac{1}{4}$ (c) $\frac{1}{6}$ (d) $\frac{1}{8}$

D2 $\frac{2}{5}$

D3 (a) $\frac{2}{6}$ or $\frac{1}{3}$ (b) $\frac{3}{8}$ (c) $\frac{5}{8}$ (d) $\frac{1}{4}$

D4 (a) $\frac{1}{6}$ (b) $\frac{2}{6}$ or $\frac{1}{3}$ (c) $\frac{3}{6}$ or $\frac{1}{2}$

D5 (a) $\frac{1}{8}$ (b) $\frac{5}{8}$ (c) $\frac{2}{8}$ or $\frac{1}{4}$

D6 $\frac{4}{5}$

D7 (a) $\frac{5}{6}$ (b) $\frac{5}{8}$ (c) $\frac{3}{8}$

D8 (a) $\frac{1}{4}$ (b) $\frac{3}{4}$ (c) $\frac{3}{8}$

What progress have you made? (p 14)

1 The game is unfair.
B wins more often than A.

2 (a) L (b) M (c) K

3 (a) $\frac{1}{5}$ (b) $\frac{4}{5}$ (c) $\frac{2}{5}$ (d) $\frac{3}{5}$

Practice booklet

Section C (p 8)

1

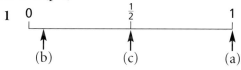

2 (a) Q (b) P (c) R

3 (a) C (b) B (c) A (d) D

Section D (p 9)

1 (a) $\frac{3}{10}$ (b) $\frac{2}{10}$ or $\frac{1}{5}$
(c) $\frac{5}{10}$ or $\frac{1}{2}$ (d) $\frac{5}{10}$ or $\frac{1}{2}$

2 (a) $\frac{3}{8}$ (b) $\frac{4}{7}$ (c) $\frac{1}{6}$ (d) $\frac{2}{9}$

3 (a) $\frac{5}{8}$ (b) $\frac{3}{7}$ (c) $\frac{5}{6}$ (d) $\frac{7}{9}$

4 Symmetry

T	p 15 **A** What is symmetrical about these shapes?	Finding out pupils' knowledge of symmetry
T	p 16 **B** Rotation symmetry	Identifying rotation symmetry Centres and orders of rotation symmetry
T	p 17 **C** Making designs	Making patterns with rotation symmetry
	p 18 **D** Rotation and reflection symmetry	Identifying rotation and reflection symmetry

---

Essential

Tracing paper
Mirrors
Square dotty paper
Triangular dotty paper
Sheets 117 to 119 and 122

Practice booklet pages 10 to 13

Optional

OHP transparencies of the shapes on
pages 15–17

A What is symmetrical about these shapes? (p 15)

Discussion should show how much pupils know already about reflection
and rotation symmetry.

Mirrors
Optional: An OHP transparency of page 15 (you may photocopy page
15 for this purpose without the normal permission), tracing paper

---

◊ One way to generate discussion is for pupils to first study the page
individually, then discuss it in small groups. The whole class can then be
brought together and contributions can be made by each of the groups.

*'Good introduction.
Many children
realise that shapes
have symmetry but
don't know why.
Discussion about
each of these shapes
helped considerably.'*

◊ Pupils should have met the reflection symmetry of the shapes before.
Some may realise that shapes with only rotation symmetry (A and H) are
symmetrical in some way but be unable to describe how. Others may
know about rotation symmetry already.

B Rotation symmetry (p 16)

> Tracing paper, sheet 117
> Optional: A transparency of the shape on the page

◊ The shape on the page is the first one on sheet 117 (shape A).

◊ In your demonstration, emphasise that the shape can be rotated by putting a pencil point at the centre of rotation and turning the tracing round this fixed point.

C Making designs (p 17)

> Tracing paper, sheets 118 and 119
> Optional: A transparency of the shape on the page

◊ In your demonstration, you may need to point out how to transfer a tracing back on to the original – either by shading on the back of the tracing paper, or by pricking through the corners of the traced shape.

The shape on the page is the first one on sheet 118 (shape A).

D Rotation and reflection symmetry (p 18)

> Tracing paper, square dotty paper, mirrors, sheet 122

B Rotation symmetry (p 16)

B1

Shape	Order of rot. symmetry
A	4
B	3
C	2
D	–
E	4
F	2
G	3
H	–
I	2
J	6
K	8
L	2
M	4

The pupil's marked centres of rotation on sheet 117

C Making designs (p 17)

C1–3 The pupil's completed designs on sheet 118

C4 The pupil's designs on sheet 119

D Rotation and reflection symmetry (p 18)

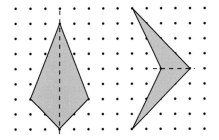

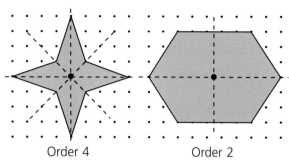

Order 4   Order 2

D2 (a) Rotation symmetry of order 2
        No reflection symmetry

(b) Rotation symmetry of order 3
        Three lines of reflection symmetry

(c) No rotation symmetry
        No reflection symmetry

(d) No rotation symmetry
        One line of reflection symmetry

(e) Rotation symmetry of order 3
        No reflection symmetry

(f) Rotation symmetry of order 2
        Two lines of reflection symmetry

(g) Rotation symmetry of order 2
        No reflection symmetry

(h) Rotation symmetry of order 2
        Two lines of reflection symmetry

(i) No rotation symmetry
        One line of reflection symmetry

(j) No rotation symmetry
        No reflection symmetry

(k) Rotation symmetry of order 2
        No reflection symmetry

(l) No rotation symmetry
        No reflection symmetry

(m) No rotation symmetry
        No reflection symmetry

(n) Rotation symmetry of order 2
        No reflection symmetry

D3 (a) Order 2

(b) Yes, it has reflection symmetry.

D4 There are sixteen ways in total to shade
four squares to make a pattern with
rotation symmetry. Pupils have to find
eight different ways.

All sixteen ways are shown below.

Rotation symmetry of order 2

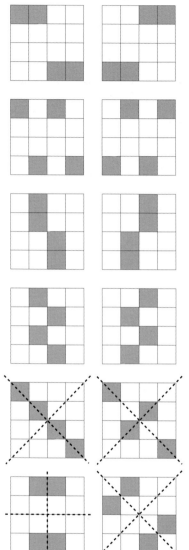

Rotation symmetry of order 4

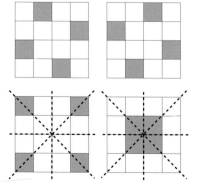

What progress have you made? (p 19)

1 (a) B, C, D and E

(b) A, B and D

(c) B and D

2, 3

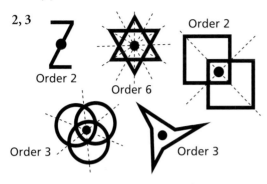

Order 2

Order 6

Order 2

Order 3

Order 3

4 The pupil's pattern; examples are

Practice booklet

Sections A and B (p 10)

1 (a) Yes (b) No (c) Yes (d) Yes

(e) No (f) No (g) Yes (h) No

(i) No

2 (a) Yes, 5 (b) Yes, 8 (c) No

(d) Yes, 4 (e) Yes, 2 (f) No

(g) Yes, 2 (h) Yes, 2 (i) No

(j) Yes, 6 (k) Yes, 2 (l) Yes, 3

Sections C and D (p 12)

1 (a) (b)

(c) (d)

(e) (f)

2 The pupil's two 5 by 5 grids with rotation symmetry

3 (a) or

(b) It is not possible.

4 The pupil's completed shapes

5 Decimals 1

This unit mainly revises work covered in *Book N* on numbers with one decimal place.

p 20	**A** One decimal place	Measuring and ordering numbers
p 22	**B** Adding and subtracting	Mental and pencil and paper methods
p 24	**C** Multiplying by a whole number	Multiplying by a single digit
p 25	**D** Place value	
p 26	**E** Multiplying and dividing by 10	

Optional
Sheet 125
Large cards with digits on them

Practice booklet pages 14 to 17

A One decimal place (p 20)

You may need to check that the basic idea of dividing units into tenths is understood. You could, for example, draw a scale on the board and ask pupils to mark given numbers on it.

B Adding and subtracting (p 22)

Both mental and written methods are included, but the emphasis is on mental methods.

◊ Draw a scale on the board marked in tenths from 0 up to 3.
Use this as a number line to show ways of adding and subtracting. For example, to do 1.7 + 0.5 you can start at 1.7, add on 0.3 to get to 2 and that leaves an extra 0.2 still to add. You can then show the process on a line on which only the whole numbers are marked, like this:

The reason for removing the tenths marks is to force a mental method, not just counting marks on a ruler.

◊ The pictures of the fish and the animals in the pupil's book are for use as the basis for oral questions, such as

   • How much does the big/small fish weigh?
   • How much do the two fish weigh altogether?
   • How much more does the big fish weigh than the small one?
   • How much more does the cat weigh than the guinea pig?

◊ Some pupils may find it helpful to replace, say, 5 by 5.0 when doing written calculations with one decimal place.

One or two (p 22)

Optional: Sheet 125

The game can be played without the cards. Players write down the nine numbers from 0.1 to 0.9 on a single piece of paper. As they choose a number, they cross it out and write it on their side of the paper.

The game can also be played by two people in front of the class, with the numbers being written on the board.

C Multiplying by a whole number (p 24)

◊ You may need to revise the basic meaning of multiplication (e.g. '3 lots of').

◊ A common error is to fail to take over 10 tenths into the units column (e.g. 1.4×3 given as 3.12). The diagram in the pupil's book shows that 3 lots of '4 tenths' = 12 tenths = 1 unit and 2 tenths. You may need to do other examples like this, using the number line.

Having dealt with examples of the type 0.3×4, you can go on to the type 2.3×4.

D Place value (p 25)

Optional: Large cards with digits on them

◊ Rule 'columns' on the board: hundreds, tens, units, tenths. Give a number, for example 267.4. Pupils holding digits make the number by standing in the right columns. You can then ask them to make a new number by, say, adding 0.1 to the number they have already. (For example, the 'tenths' pupil holding 4 would be replaced by another holding 5.)

◊ The same set-up can be used to show multiplication and division by 10 (see section E).

E Multiplying and dividing by 10 (p 26)

If pupils have learned a rule 'to multiply by 10, add a nought', they will find it breaks down where decimals are involved.

> Optional: Large cards with digits on them

◊ Before, or instead of, the calculator-based activities, you could use 'pupil digits' as described in the previous section.

When a number (e.g. 62.3) has been made, you can ask for it to be multiplied by 10. Pupils should know that the 60 becomes 600 and the 2 becomes 20, so you can focus attention on what happens to the 0.3.

Dividing by 10 can be approached in a similar way, as the inverse of multiplying by 10.

A One decimal place (p 20)

A1 (a) 10.8 cm (b) 6.3 cm (c) 0.7 cm

A2 (a) 4.2 cm (b) 4.2 cm
(c) 4.2 cm (d) 4.2 cm

A3 (a) The pupil's judgement
(b) WX = 4.4 cm, YZ = 4.6 cm
(c) YZ

A4 PQ = 5.6 cm, RS = 5.8 cm, TU = 5.7 cm

A5 $r = 3.7$ cm, $p = 5.1$ cm,
$s = 5.3$ cm, $q = 7.1$ cm

A6 (a) 0.4 (b) 1.6 (c) 2.2
(d) 3.1 (e) 4.5

A7 (a) 0.7 litre (b) 0.1 litre

A8 3.5

A9 (a) 6.5, 6.9, 7, 7.2, 7.6
(b) 3.9, 4, 4.2, 5.3, 6
(c) 0.6, 0.8, 1, 1.1, 1.2, 2.4
(d) 0.7, 1.3, 2.6, 2.8, 3, 5

B Adding and subtracting (p 23)

B1 (a) 1.1 (b) 1.2 (c) 1.9
(d) 4.5 (e) 4(.0)

B2 (a) 1.4 (b) 0.8 (c) 0.8
(d) 1.7 (e) 2.6

B3 Mel

B4 Sam

B5 (a) 13.4 kg (b) 9.2 kg (c) 1.6 kg
(d) 2.6 kg (e) 4.2 kg

B6 (a) 12.8 (b) 10.3 (c) 8.2
(d) 11.5 (e) 14.4

B7 (a) 3.3 (b) 4.5 (c) 5.4
(d) 5.3 (e) 5.8

B8 (a) 2.9 (b) 6(.0) (c) 10.3
(d) 3.8 (e) 15.5

C Multiplying by a whole number (p 24)

C1 (a) 0.8 (b) 1.2 (c) 1.5
(d) 1.6 (e) 2.0 (or 2) (f) 2.4

C2 7.5 litres

C3 (a) 9.6 (b) 30.1 (c) 16.8
(d) 4.2 (e) 21.6 (f) 29.6

C4 (a) 4.2 kg (b) 13 kg or 13.0 kg
(c) 20.4 kg (d) 22.4 kg

C5 20.8 metres

C6 $28.4 \times 7 = 198.8$
The truck can carry all the boxes.

Ⓓ Place value (p 25)

D1 (a) 1 thousand (1000)
    (b) 6 hundreds (600)
    (c) 4 tens (40)
    (d) 5 units (5)

D2 (a) 4 tens (40)   (b) 9 tenths (0.9)
    (c) 2 hundreds (200)

D3 (a) 46.6   (b) 55.6   (c) 45.7
    (d) 135.8   (e) 136.7   (f) 145.7

D4 (a) 33.8   (b) 42.8   (c) 32.9
    (d) 147.4   (e) 137.5   (f) 138.4

D5 50

D6 7

D7 0.7

Ⓔ Multiplying and dividing by 10 (p 26)

E1 (a) 34   (b) 2   (c) 14
    (d) 21   (e) 3   (f) 17
    (g) 54   (h) 151   (i) 27
    (j) 204

E2 (a) 137   (b) 532   (c) 18
    (d) 46   (e) 56   (f) 33
    (g) 7   (h) 194   (i) 325
    (j) 160

E3 (a) 4.3   (b) 2.7   (c) 13.3
    (d) 64.8   (e) 0.6   (f) 6.7
    (g) 2.5   (h) 0.4   (i) 4.2
    (j) 12.5

E4 (a) 3.6   (b) 0.9   (c) 1.6
    (d) 7 or 7.0   (e) 3.1   (f) 13.6
    (g) 5.5   (h) 13   (i) 2.1
    (j) 0.7

E5 (a) 47   (b) 3.6   (c) 380
    (d) 1   (e) 26.8   (f) 15.7
    (g) 386   (h) 7.4   (i) 710
    (j) 1.9

E6 (a) 4.5   (b) 670   (c) 22
    (d) 60.3   (e) 2   (f) 650
    (g) 32   (h) 4   (i) 12.1
    (j) 223

What progress have you made? (p 27)

1 (a) 5.6   (b) 5.3   (c) 3.3
    (d) 11.3   (e) 6.8   (f) 2.9

2 (a) 3 or 3.0   (b) 10.8

3 (a) 7 tens (70)   (b) 3 tenths (0.3)

4 (a) 56   (b) 36.2   (c) 23.1
    (d) 43   (e) 0.8   (f) 7

Practice booklet

Section A (p 14)

1 (a) (i) 3.3 cm   (ii) 3.6 cm
      (iii) 3.4 cm   (iv) 3.5 cm
  (b) No, the sides are not the same
     length.

2 (a) The pupil's estimates
  (b) Line *a*: 3.6 cm
     Line *b*: 3.9 cm
     Line *c*: 4.1 cm
     Line *d*: 3.8 cm
     Line *e*: 3.5 cm
     The longest line is line *c* (4.1 cm).
     The shortest line is line *e* (3.5 cm).

3 (a) 0.2   (b) 1.1   (c) 1.9
    (d) 2.3   (e) 7.5   (f) 9.5

4 (a) 1.2, 1.5, 1.9, 2, 2.6
  (b) 5.9, 6, 6.2, 7.3, 8
  (c) 0.3, 0.7, 1, 1.2, 1.4, 2.7
  (d) 0.6, 2, 6.8, 7.3, 11

Section B (p 15)

1 (a) 0.7 litre   (b) 0.4 litre
   (c) 1.1 litres

2 (a) 0.6 kg   (b) 4.4 kg

3 (a) 7.7   (b) 4.3   (c) 3.9
   (d) 18.7   (e) 1.2
   (f) 2 (or 2.0) (g) 7.2   (h) 5.5

4 (a) 6.1   (b) 7 or 7.0
   (c) 5.9   (d) 2.5   (e) 0.4
   (f) 3.4   (g) 0.3   (h) 2.8

5 (a) 0.1 and 0.8 or
        0.2 and 0.7
   (b) 0.9 and 0.5
   (c) 0.1 and 0.9 or
        0.2 and 0.8 or
        0.3 and 0.7
   (d) 0.8, 0.9 and 0.7
   (e) 0.8, 0.5 and 0.7 or
        0.8, 0.9 and 0.3
   (f) 0.2, 0.1, 0.8, 0.9, 0.3 and 0.7 or
        0.1, 0.8, 0.9, 0.5 and 0.7

Section C (p 16)

1 6 kg

2 4.2 kg

3 (a) 2.4   (b) 3 or 3.0   (c) 4 or 4.0
   (d) 3.5   (e) 0.4   (f) 1.5
   (g) 1.8   (h) 2.1

4 (a) 6.8   (b) 12.6   (c) 11.2
   (d) 36.4   (e) 5.4   (f) 51.2
   (g) 5.4   (h) 56.7

5 20.4 metres

6 6.3 metres

7 Yes, the lift can carry the boxes.
   $46.2 \times 8 = 369.6$
   369.6 kg is less than 370 kg so it is just
   possible.

Sections D and E (p 17)

1 (a) 5 tens (50)
   (b) 8 tenths (0.8)
   (c) 6 hundreds (600)

2 (a) 33.7   (b) 32.8   (c) 42.7
   (d) 106.5   (e) 116.4   (f) 107.4

3 (a) 48.6   (b) 57.6   (c) 47.7
   (d) 284.5   (e) 275.5   (f) 274.6

4 (a) 68   (b) 6   (c) 18
   (d) 315   (e) 45   (f) 2
   (g) 206   (h) 11

5 (a) 7.5   (b) 1.6   (c) 0.8
   (d) 14.7   (e) 0.4   (f) 2
   (g) 0.3   (h) 9.9

6 (a) 490   (b) 4.9   (c) 60.3
   (d) 106   (e) 52.1   (f) 3.2
   (g) 27   (h) 19

Review 1 (p 28)

1 (a) 8:30 a.m. (b) 2:15 a.m.
(c) 5:45 a.m. (d) 9:45 p.m.
(e) 10:40 p.m. (f) 6:35 a.m.
(g) 7:55 p.m. (h) 3:05 a.m.

2 (a) 5:15 p.m. (b) 11:30 p.m.
(c) 6:45 a.m. (d) 5:50 p.m.

3 (a) 20 minutes (b) 1 hour 45 minutes
(c) 25 minutes (d) half past 12

4 (a) $\frac{2}{4}$ or $\frac{1}{2}$ (b) $\frac{2}{3}$
(c) $\frac{4}{6}$ or $\frac{2}{3}$ (d) $\frac{2}{5}$

5 (a) $\frac{2}{4}$ or $\frac{1}{2}$ (b) $\frac{1}{3}$
(c) $\frac{2}{6}$ or $\frac{1}{3}$ (d) $\frac{3}{5}$

6 (a) Yes, order 2 (b) No
(c) Yes, order 2 (d) Yes, order 4
(e) No

7 (a) 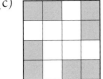 (b)

(c)

8 (a) 0.8 (b) 1.3 (c) 2.6
(d) 3.5 (e) 4.6

9 (a) 0.9, 1.1, 4.9, 5, 8
(b) 7.5, 8.7, 9.1, 9.9, 10
(c) 0.2, 0.7, 0.8, 1, 1.1
(d) 2.9, 3, 3.5, 5.5, 6

10 (a) 3.5 (b) 3.4 (c) 4.1 (d) 5.8
(e) 1.7 (f) 1.2 (g) 3.8 (h) 4.6
(i) 4.8 (j) 4.5

11 (a) 2.6 (b) 4.8 (c) 3.6 (d) 3.2
(e) 10.4 (f) 19 (g) 25 (h) 150
(i) 273 (j) 183 (k) 5.5 (l) 0.9
(m) 10 (n) 1.7 (o) 6.5

Mixed questions 1 (practice booklet p 18)

1 (a) 45 minutes (b) 35 minutes
(c) 2 hours 35 minutes
(d) 1 hour 25 minutes

2 (a) 5 past 9 in the morning or 9:05 a.m.
(b) 10 past 3 in the afternoon or 3:10 p.m.
(c) 5 past 6 or 6:05
(d) 8:50 p.m. or 10 to 9 at night

3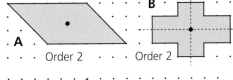

4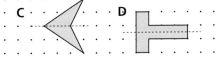

5 (a) 1.6 (b) 9.9 (c) 1.7 (d) 2.2
(e) 5.8 (f) 5.5 (g) 6.7 (h) 0.3
(i) 7 (j) 3.4 (k) 206 (l) 10.1

⑥ Number grids

In this unit, pupils solve number grid problems using addition and subtraction. This includes using the idea of an inverse operation ('working backwards').

There are opportunities to work investigatively, exploring and describing number patterns.

Algebra is introduced in the context of number grids. Pupils simplify expressions such as $n + 4 + 3$.

T	p 30	**A** Square grids	Simple addition and subtraction problems
			Investigating 'diagonal rules' on number grids
	p 32	**B** Grid puzzles	Solving addition and subtraction problems where the appropriate operation must be chosen
			Using the idea of an inverse operation ('working backwards') to solve number puzzles
			Solving more difficult puzzles, possibly by trial and improvement
	p 35	**C** Grid investigations	Exploring and describing number patterns
T	p 37	**D** Algebra on grids	Knowing that, for example, a number 2 more than n is $n + 2$
			Simplifying expressions such as $n + 4 - 3$

> **Optional**
> Sheets 126 to 129
> A4 sheets of paper and felt-tip pens or crayons
> Squared paper
>
> **Practice booklet** pages 19 and 20

Ⓐ Square grids (p 30)

The idea of a number grid is introduced. There are many opportunities to discuss mental methods of addition and subtraction.

> Optional: Sheet 126 may be useful for the activities on page 30 and question A1; squared paper is useful for drawing grids; A4 sheets of paper and felt-tip pens or crayons (for *Human number grids*)

Human number grids

This introductory activity does not appear in the pupil material.

◊ Each pupil or pair of pupils represents a position in a number grid. (Number grids are on page 30 of the pupil's book.)

Each position will contain a number. (For pupils familiar with spreadsheets, the idea of a 'cell' may help.)

The operations used are restricted to addition and subtraction.

◊ Tables/desks need to be arranged in rows and columns so that the cells form a grid. Explain, with appropriate diagrams, that the class is going to form a human number grid that uses rules to get from a number in one cell to a number in another. A possible diagram is shown below.

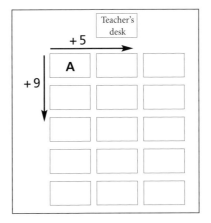

◊ Referring to the numbers is easier if the cells are labelled.
Pupils can discuss how each cell might be labelled, for example:

• A1, A2, B1, … as on a spreadsheet

• A, B, C, …

• or with the pupils' names

◊ Initially, it may be beneficial to use only addition or use sufficiently large numbers in cell A to avoid the complication of negative numbers.

◊ Decide on the first pair of rules and ask the pupils in the cell marked A in the diagram to choose a number for that cell.
Discuss how the numbers in other cells are found.
Now ask the pupils in cell A to choose another number, write it on both sides of a sheet of paper and hold it up.
Pupils now work out what number would be in their cell, write it on both sides of their sheet of paper and hold it up.
This can be repeated with different pupils deciding on the number for their cell.

'Human number grids was a very helpful introduction – returned to it several times throughout unit.'

'This activity was an effective introduction to the idea of a number grid but was a bit of a nightmare last thing in the afternoon!'

'Need to ensure that lower attainers are not sitting in places at the bottom right!'

◊ Questions can be posed in a class discussion, for example:
  - Suppose the number in Julie and Asif's cell is 20.
    What number is in your cell, Peter?
    What number will be in Jenny's cell?
  - What number do we need to put in cell A so that
    the number in cell F is 100?
  - What happens if the 'across' and 'down' rules change places?
  Ask pupils to explain how they worked out their answers.

Square grids (p 30)

◊ Point out that all grids in the unit are square grids.

◊ One teacher presented unfinished grids on an OHP transparency and
  asked for volunteers to fill in any empty square. She found that less
  confident pupils chose easy squares to fill in, while 'others with more
  confidence chose the hardest, leading to class discussion. The idea of a
  "diagonal" rule came out naturally.'

◊ In one school, the class looked at rules in every
  possible direction as shown in the diagram.

◊ In discussion, bring out the fact that there are
  different ways to calculate a number in a square
  depending on your route through the grid.
  For example, the number in the bottom
  right-hand square in a 3 by 3 grid can be reached in six different ways.
  Pupils can try to find all these ways.
  Investigating the number of different routes through a grid to each square
  can lead to work on Pascal's triangle.

'Many were surprised by the fact that there was more than one route from one square to another, giving the same answer.'

◊ Make sure pupils realise that the diagonal rule fits any position on these
  grids and not just those on the leading diagonal.

A1 In part (b), emphasise that the diagonal rule fits any position on the grid.

A2 It is likely that negative numbers will appear in the grids, providing an
   opportunity to consolidate work on negative numbers. However, if you
   want to try to avoid this, you could suggest that pupils choose quite large
   numbers for the top left-hand square of their grids or stick to rules that
   involve addition only.

   Appropriate teacher input is important here. Pupils could
   - consider rules that involve addition only
   - choose two rules and investigate grids that use those rules only
   - consider rectangular grids

A3 Some pupils may need to enter numbers in the grid to find the diagonal
   rules. Encourage pupils to use the results of their investigation in A2 to
   calculate the diagonal rules from the across and down rules only.

B Grid puzzles (p 32)

Pupils solve grid puzzles where the appropriate operation (addition or subtraction) must be chosen. They use the idea of 'working backwards'.

Some extension material is included in which pupils could use trial and improvement to solve number grid puzzles.

Optional: Sheets 127 to 129 (to record solutions)

For B1 to B3, emphasise that it is not necessary to complete the whole grid to solve the puzzle, just to find the missing number or rules.

B4 As a possible extension, pupils could make up their own puzzles like this to solve. However, puzzles without at least one pair of numbers in the same horizontal or vertical row are easy to construct but more difficult to solve. These may provide an enjoyable challenge, but if they lead to frustration, pupils could be restricted to making up puzzles with at least one pair of numbers in the same horizontal or vertical row.

B5 This is similar to question A4. Emphasise that this shows that not all puzzles with missing rules have a unique solution.

*B6 This puzzle is likely to be much more difficult than earlier ones. Pupils can solve it using trial and improvement.

*B7 Pupils can devise their own trial and improvement methods to solve these. Encourage them to be systematic and ask them to explain their methods to you or each other.

*B8 Encourage pupils to be systematic in their choice of pairs to make the link easier to find. The link could be expressed in words: for example, 'If the across rule is to add a number, then the down rule is to subtract twice that number'.

C Grid investigations (p 35)

Optional: Squared paper may be useful for pupils to draw grids on.

As an introductory activity, pupils could consider grids where the across and down rules are exactly the same.

C3 Make sure that pupils are aware that a 'diagonal' in this case must go from corner to corner. Some pupils may confuse the 'diagonal rule' which holds all over a grid with a 'diagonal' which goes from corner to corner.

C4 Pupils are likely to need support with this question.

Ⓓ **Algebra on grids** (p 37)

Algebra is introduced in the context of number grids.

◊ In the teacher-led introduction, you may find a number line is helpful in getting these ideas across.

Remind pupils of earlier work in section A on finding rules. Emphasise that the expressions in the grid show how to **directly** find any number on the grid from the top left-hand corner. For example, as the expression in the bottom right corner is $n + 14$, then the rule to go from the number in the top left square to the number in the bottom right square is '+ 14'.

Ⓐ **Square grids** (p 31)

A1 (a)

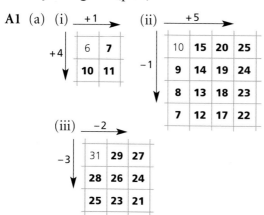

(b) (i) + 5 (ii) + 4 (iii) – 5

A2 The pupil's investigation and description of the link

A3 (a) + 17 (b) + 5

A4 The pupil's pairs of rules, for example,

(a) across '+ 1', down '+ 10', across '+ 5', down '+ 6'

(b) across '+ 1', down '+ 3', across '– 1', down '+ 5'

Ⓑ **Grid puzzles** (p 32)

B1 (a) 17 (b) 23* (c) 70* (d) 39

(e) 20 (f) 15 (g) 25 (h) 19

(i) 8

*Pupils who have not grasped the idea that the rules operate from left to right and from top to bottom might give 27 and 130 as their answers for parts (b) and (c) respectively.

B2 (a) + 4 (b) + 2 (c) – 4

B3 The pupil's puzzles

B4 (a) Across '+ 4', down '+ 9'

(b) Across '– 4', down '– 2'

(c) Across '– 3', down '+ 5'

(d) Across '+ 3', down '+ 8'

(e) Across '+ 7', down '+ 3'

(f) Across '– 5', down '– 6'

(g) Across '+ 2', down '+ 3'

(h) Across '– 2', down '+ 11'

(i) Across '– 5', down '– 3'

B5 The pupil's pairs of rules, for example,

across '+ 1', down '+ 9'
across '+ 2', down '+ 8'

across '+ 3', down '+ 7'
across '+ 4', down '+ 6'

across '– 1', down '+ 11'
across '+ 20', down '– 10'

*B6 (a) Across '+ 2', down '– 3'

(b) The pupil's reasons

*B7 (a) Across '+ 1', down '+ 2'

(b) Across '– 1', down '+ 5'

(c) Across '– 3', down '– 4'

(d) Across '+ 4', down '– 1'

(e) Across '+ 6', down '– 2'

*B8 (a) The pupil's pairs of rules, for example,

across '+ 1', down '– 2'
across '+ 2', down '– 4'

across '+ 3', down '– 6'
across '– 4', down '+ 8'

(b) Pupil's descriptions of the link, for example,

'If the across rule is to add a number, then the down rule is to subtract twice that number';

'If the across rule is to subtract a number, then the down rule is to add twice that number'.

C **Grid investigations** (p 35)

C1 (a) (i)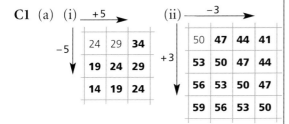

(iii)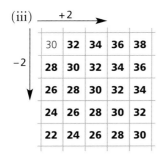

(b) The pupil's investigations

(c) The pupil's observations about symmetry and diagonals

C2 (a) $37 + 7 = 44$

(b) (i) Each pair adds to give 46.

(ii) Each pair adds to give 179.

(c) The pupil's grids

(d) The opposite corners' totals are the same each time.

C3 (a) $25 + 32 + 39 + 46 = 142$
$37 + 36 + 35 + 34 = 142$

(b) (i) The numbers in each diagonal add to give 48.

(ii) The numbers in each diagonal add to give 184.

(c) The pupil's grids

(d) The diagonal totals are the same each time.

C4 (a)

Diagonals table	
Top left number	Diagonal total
2	9
3	12
4	15
10	33

(b) The pupil's grids

(c) The pupil's results in table

(d) 'The diagonal total is 3 times the top left number add 3' or
'The diagonal total is (the top left number × 3) + 3' or
'The diagonal total is (the top left number + 1) × 3' or equivalent.

D **Algebra on grids** (p 37)

D1 (a)

$$+1$$

n	$n+1$	$n+2$
$n+4$	$n+5$	$n+6$
$n+8$	$n+9$	$n+10$

$+4$

(b) 110

(c) 40

D2 (a)
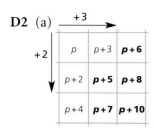

(b)
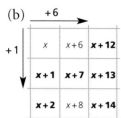

(c)
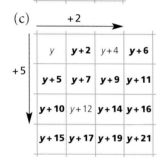

(d)
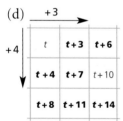

(e)
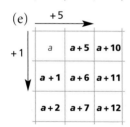

D3 (a) 20   (b) 24   (c) 31

      (d) 24   (e) 22

D4 (a) Across '+ 1', down '+ 2'

      (b) Across '+ 6', down '+ 5'

      (c) Across '+ 4', down '+ 1'

      (d) Across '+5', down '+ 1'

      (e) Across '+ 3', down '+ 10'

D5 (a) $f + 8$ (b) $y + 8$

      (c) $x + 3$ (d) $z + 11$

      (e) $p + 6$ (f) $m + 9$

What progress have you made? (p 39)

1 (a)   (b)
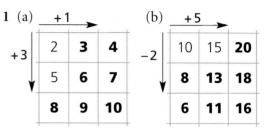

2 (a) 4   (b) 11

3 (a) Down '+ 2'   (b) Across '+ 5'

     (c) Across '+ 6', down '– 2'

     (d) Across '– 5', down '– 3'

4 (a)   (b)
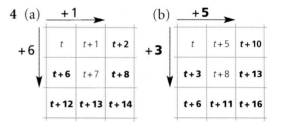

Practice booklet

Section A (p 19)

1 (a)
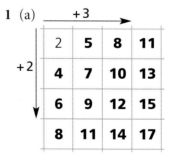

(b)
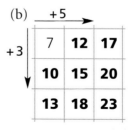

(c)

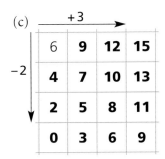

(d)

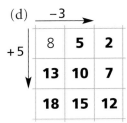

(e)

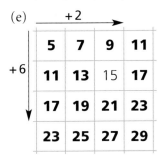

(f)

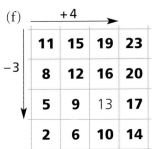

2 (a) + 5   (b) + 8   (c) + 1
   (d) + 2   (e) + 8   (f) + 1

Section B (p 19)

1 (a) + 7   (b) + 2   (c) + 10

2 The pupil's pairs of rules, for example,

(a) across '+ 1', down '+ 5'
across '+ 2', down '+ 4'

across '+ 3', down '+ 3'
across '+ 7', down '– 1'

(b) across '+ 2', down '– 2'
across '+ 1', down '– 1'

across '– 3', down '+ 3'
across '+ 4', down '– 4'

Section D (p 20)

1 (a) (i)

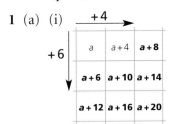

(ii)

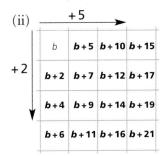

(b) (i) 40   (ii) 41

(c) (i) 10   (ii) 9

2 (a) Across '+ 3', down '+ 1'

(b) Across '+ 5', down '+ 3'

(c) Across '+ 2', down '+ 4'

3 (a) $t + 11$ (b) $a + 7$ (c) $q + 12$
   (d) $p + 3$ (e) $x + 6$ (f) $y + 8$
   (g) $s + 13$ (h) $v + 19$ (i) $b + 13$
   (j) $a + 15$ (k) $f + 21$ (l) $c + 18$
   (m) $d + 10$ (n) $g + 16$ (o) $h + 14$

Stars and angles

The purpose of the unit is for pupils to practise measuring and estimating angles.

| T | p 40 | **A** Five-pointed stars | Drawing stars and measuring their angles |
| T | p 40 | **B** Estimating angles | |

Essential

Plain paper (several sheets per pupil)
Angle measurers (one per pupil)
Sheets 130 to 133

A Five-pointed stars (p 40)

> Plain paper, angle measurers

While the main purpose of this section is to give pupils practice in measuring angles, it also provides an opportunity for them to make a generalisation from their own results. There will inevitably be errors in their measurements, but with guidance pupils will be able to see that the angles at the points of the stars add up to 180°.

Extension

'We moved on to 6-pointed stars (by accident!) and then 7. Went really well!'

When pupils have mastered the technique of drawing the stars and concluded that the total of the angles at the points is 180° they could be challenged to try to draw some 'regular' stars (with each angle 36° and all the lines the same length).

B Estimating angles (p 40)

> Sheets 130 to 133, angle measurers

The sheets are designed to be used in pairs, 130 and 131, 132 and 133.

'I got a pupil to draw an angle on the board. Everyone estimated it, and another pupil measured it. The closest pupil drew the next angle.'

1 Pupils estimate all the angles on the first sheet in the pair.
2 They check with an angle measurer, and work out their 'total error'.
3 You may then wish to discuss methods of estimating the angles.
4 Now pupils repeat steps 1 and 2 with the second sheet, and see if their estimation skills have improved.

Sheets 132 and 133 include obtuse angles.

◊ When discussing methods of estimating angles, you might include:
  • turning the sheet until one arm of the angle is horizontal
  • drawing by eye a 90° line   • drawing by eye a 45° line

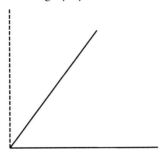   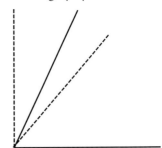

There are a number of software packages available that allow pupils to practise their angle estimation skills.

B **Estimating angles**

Pupils' answers will inevitably differ slightly from these.

	130	131	132	133
1	25°	76°	19°	142°
2	55°	28°	120°	15°
3	10°	52°	90°	163°
4	35°	12°	78°	98°
5	80°	38°	155°	82°
6	63°	66°	170°	100°
7	18°	9°	32°	125°
8	48°	85°	96°	39°
9	87°	27°	111°	152°
10	7°	44°	137°	61°

Lines at right angles

For many pupils, this short unit will be a piece of revision.

p 41   **A** Thinking about right angles   Using the fact that a right angle is 90°

p 42   **B** Drawing and checking right angles   Developing accuracy

---

Essential

Set square
Plain paper

Practice booklet pages 21 and 22

Thinking about right angles (p 41)

T

'They were astonished how many different right angles came up.'

Pupils may need reminding of what a right angle is, and how many degrees make up a right angle. The photographs on page 41 contrast one building consisting entirely of right angles and another where they are very hard to find. You could then ask pupils to each list 20 right angles in their classroom. Their lists can then be discussed.

A2 Some pupils may be uncertain about the points of the compass and find it difficult to relate them to turning; clockwise and anticlockwise may also be a problem. If so, it's worth developing question A2 into a class activity.

First establish the direction of north in relation to your classroom. Then have pupils take turns to stand facing in a given direction, follow instructions to turn clockwise or anticlockwise through a right angle and say what direction they are now facing.

Drawing and checking right angles (p 42)

These simple exercises aim to develop accuracy. Pupils with poor hand–eye skills should be given sufficient time and the chance to have a second 'go' when first attempts go awry.

Plain paper is essential for these questions

A Thinking about right angles (p 41)

A1 12 and 3, 1 and 4, 2 and 5, 3 and 6,
4 and 7, 6 and 9, 7 and 10, 8 and 11,
9 and 12, 10 and 1, 11 and 2

A2 (a) North   (b) North-east

(c) North-east

B Drawing and checking right angles (p 42)

B1–B4 The pupil's drawings

B5 a and b, c and j, d and e, i and l

B6 Most people think the eight angles at the
ends of the arms are right angles but they
are not. The other four angles look too
large or too small but they are in fact
right angles.

What progress have you made? (p 44)

1 a and b, c and d, e and f

2 a and e, b and h, c and f, d and g

3 The pupil's drawing

Practice booklet

Sections A and B (p 21)

1 90 (degrees)

2 (a) 5   (b) 11

3 (a) East   (b) West

(c) South-west   (d) North-west

(e) South-east

4 The pupil's drawings

5 a and e, b and j, c and i, d and f, h and k.
g is the odd one out.

Comparisons

This unit introduces median and range and uses these to compare groups for different sets of data.

Pupils generate their own data in a reaction time experiment and memory tests and the data can be used to make comparisons.

p 45	**A** Median	Finding the median from an odd number of items
p 47	**B** Dot plots	Finding medians from dot plots
p 48	**C** Middle pairs	Finding the median of an even number of items
p 49	**D** Comparing groups	Using the median and range for comparisons
p 51	**E** How fast do you react?	Making and using a simple reaction timer to generate data for comparison
p 52	**F** The Argon Factor	Handling data from mental exercises

---

Essential

Sheets 140 to 144
Graph paper

Practice booklet pages 23 to 26

A Median (p 45)

Median can be introduced by using groups of pupils themselves as 'equipment'.

◊ Pupils will need to know their heights in centimetres or will need to measure them.

Put an odd number of pupils in a line in order of height. Emphasise that it is the height of the middle person (not the person himself or herself) that is the median height of the group. Using the word 'median' as an adjective, as in 'median height' helps to emphasise the units of the median and makes it less likely that the median will be thought of as a person.

The heights of the Victorian children can be read from the scales at the sides of the picture. Their median height can be found and compared with the class.

'This was good – using decimals when the concept was embedded.'

◊ If pupils ask what happens with an even number of pupils, you can deal with that here. Otherwise it can be left until section C.

◊ Numbers with one decimal place appear in some questions. You may want to revise this first.

B **Dot plots** (p 47)

Sheet 140

A dot plot is a one-dimensional scatter diagram and very useful for getting an overall picture of a small data set. Generally dots are placed above the line but when there are two separate sets of data these are usefully placed either side of the line.

C **Middle pairs** (p 48)

Sheet 140

◊ You could introduce this as in section A, with a group of pupils.

◊ Adding the middle pair of heights together and dividing by 2 can seem a mysterious way of finding the height halfway between. Pupils may be happier finding the difference, halving it and adding that to the lower height.

D **Comparing groups** (p 49)

The idea that there are two measures, 'average' and 'spread', which can be used to compare groups should be emphasised in this section.

◊ The range can be shown practically with a group of pupils. Ask the group to stand in order of height. Pick out the tallest and shortest and ask them to stand next to one another. Measure the difference between their heights. (This can often be done with an ordinary 30 cm ruler.) The difference is the range of the heights of the group.

◊ Emphasise that the range is a single value, i.e. biggest minus smallest, and has units.

E **How fast do you react?** (p 51)

Pupils work in pairs.

Sheet 141

◊ In addition to comparing performance within each pair, pupils could compare left and right hands. For homework they could compare themselves with an adult.

◊ In one class pupils felt that they were getting clues from twitching fingers just before the ruler was dropped, so a card was used to cover the fingers.

◊ In the class as a whole, comparisons could be made between boys and girls.

F **The Argon Factor** (p 52)

Sheets 142 and 143 (preferably on OHP transparencies) Sheet 144

◊ There are two tests: mental agility and memory.

◊ In the *mental agility* test, give the pupils one minute to memorise the shapes and numbers. (They are best shown on an OHP.)
Tell pupils 'You will be given 5 seconds to answer each question. Questions will be read twice.'

1   What number is inside the circle? (21)
2   What number is inside the pentagon? (33)
3   What number is inside the first shape? (13)
4   What shape has the number 17 in it? (triangle)
5   What shape has the number 29 in it? (kite)
6   What number is inside the middle shape? (17)
7   What number is inside the last shape? (33)
8   What shape is in the middle? (triangle)
9   What is the shape before the end one? (kite)
10   What number is in the second shape? (21)
11   What shape is the one after the one with 21? (triangle)
12   What number is inside the fourth shape? (29)
13   What number is in the shape just right of the triangle? (29)
14   What shape is two to the left of the kite? (circle)
15   What number is in the shape two after the circle? (29)

◊ In the *memory test*, give the pupils 2 minutes to remember the pictures and details of the four people. (You might want to use fewer pictures or less information with some classes.) Details and questions may need to be read out if pupils have reading difficulties.

Then give them 10 minutes to answer the 20 questions on paper.
The answers to the memory tests are:

1	12	**11**	Maths
2	Brown	**12**	Football
3	Canoeing	**13**	10
4	Blue	**14**	Art
5	Miss Ward's	**15**	Stripes
6	Bird watching	**16**	Flowers
7	Brown	**17**	3
8	12	**18**	Sarah
9	Music	**19**	Mital
10	Mr Wright's	**20**	No one

◊ Discuss with the class what comparisons can be made from the scores. Suggestions:

• Do people remember more about their own sex?
• Which test did the class do better on? (Remember that the number of questions is not the same.)

Ⓐ **Median** (p 46)

A1 (a) 147, 154, 156, 163, 164, 169, 180 cm
 (b) 163 cm

A2 (a) 23, 29, 37, 44, 50 kg
 (b) 37 kg

A3 (a) 61 kg (weights in order are 52, 57, 58, 61, 66, 70, 76 kg)
 (b) 96 kg (weights in order are 89, 94, 96, 103, 107 kg)
 (c) 128 kg (weights in order are 117, 120, 124, 126, 128, 130, 134, 145, 148 kg)

A4 (a) 13, 13, 16, 17, 23, 25, 34
 (b) 17

A5 (a) 16   (b) 15   (c) 16

A6 6.2 m

Ⓑ **Dot plots** (p 47)

B1 (a) 159 cm   (b) 157 cm
 (c) 154 cm   (d) 154 cm

B2 Medians:
 (a) 126 cm   (b) 97 cm
 (c) 8.8 kg   (d) 20.3 kg
 (It may be useful to mark the answers on a copy of sheet 140.)

B3 (a) 156 cm
 (b) 156 cm
 (c) 158 cm

Ⓒ **Middle pairs** (p 48)

C1 (a) 157 cm (b) 155 cm (c) 154 cm

C2 Medians:
(a) 141 cm (b) 83 cm
(c) 74 cm (d) 11.5 cm
(It may be useful to mark the answers on a copy of sheet 140.)

C3 (a) 143 cm (b) 148 cm

Ⓓ **Comparing groups** (p 49)

D1 (a) 130 cm (b) 140 cm (c) Boys

D2 (a) Girls: 14 cm Boys: 18 cm
(b) Boys

D3 15 cm

D4 4 cm (9 − 5 = 4)

D5 (a) Team A: 70 kg Team B: 73 kg
(b) Team B
(c) Team A: 18 kg Team B: 23 kg
(d) Team B

D6 (a) True (b) False (c) False (d) True

What progress have you made? (p 52)

1 35 **2** 12

3 19 **4** 14

5 Left hand: median 18, range 16
Right hand: median 15, range 8
Sunil is faster and less variable using his right hand.

Practice booklet

Section A (p 23)

1 (a) 109, 117, 125, 130, 131 cm
(b) 125 cm

2 (a) 24, 27, 31, 33, 36, 36, 39 kg
(b) 33 kg

3 (a) 97 kg (b) 55 kg (c) 13
(d) 17°C (e) 48 (f) 8

Sections B and C (p 24)

1 128 cm **2** 128 cm

3 (a) 138 g (b) 140 g (c) 207 g

4 (a)

```
  ├┼┼┼┼┼┼┼┼┼┼┼┼┼┼┼┼┼┼┼┼┼┼┤
 110      120      130      140      150 gram
         • •           •      •         • •
```

(b) 132 g

Section D (p 25)

1 (a) Median: 61 kg, range: 21 kg
(b) Median: 3.9 kg, range: 1.7 kg
(c) Median: 13 cm, range: 9 cm
(d) Median: 9.7 m, range: 10.8 m
(e) Median: 7°C, range: 9 degrees

2 (a) Tosca 68 g Murdo 70 g
(b) Tosca 22 g Murdo 20 g
(c) Murdo

3 (a) Barcelona: 21°C, Birmingham: 22°C
(b) Birmingham
(c) Barcelona: 12 degrees
Birmingham: 8 degrees

4 (Lengths are A: 27 mm, B: 33 mm, C: 36 mm, D: 40 mm, E: 56 mm, F: 53 mm.)
Median length 38 mm
Range of lengths 29 mm

5 (a) True, red median weight is 294 g and grey median weight is 596 g.
(b) True, range for red squirrels is 25 g and for grey squirrels is 112 g.

6 (a) Biddy: median 58 g, range 8 g
Mabel: median 60 g, range 26 g
(b) (i) True (ii) True

7 (a) North: median 12 m, range 7 m
South: median 14.5 m, range 12 m
(b) Trees on the south side are taller. They are more spread out in height.

Ice cream (p 53)

In this unit pupils simulate running an ice cream stall. They have to order stock on the basis of temperature predictions. The actual temperature may be above or below the prediction. They can then work out how much profit they have made.

This activity involves a lot of mental arithmetic and organisational skills.

Essential	Optional
Sheet 146 (one per pair)	Sheet 147 (harder version)
One dice	

This activity should be teacher-led, with pupils working in pairs, each pair running a stall.

◊ Each pair starts with the same amount of money (e.g. £10).
They record this in the Monday 'Cash in box at start of day' cell.

◊ You throw the dice and announce the temperature forecast for Monday, using these values.
  1 or 2: 10°C   3 or 4: 15°C   5 or 6: 20°C

◊ Each pair now has to decide how many ice creams (at 40p each) to order. Explain that the actual temperature could be above or below the forecast, as follows.

  Forecast: 10°C   Actual will be 7°C or 12°C
  Forecast: 15°C   Actual will be 12°C or 17°C
  Forecast: 20°C   Actual will be 17°C or 22°C

The number sold will depend on the actual temperature:
  7°C   0 sold   12°C   10 sold
  17°C   15 sold   22°C   30 sold

When pupils have decided how many ices to order, they fill in the order on the sheet and work out the cash left in the box.

◊ Now you throw the dice again. If it lands odd, you go down from the forecast to get the actual temperature (e.g. a 15°C forecast becomes 12°C actual); if even, you go up.

◊ Each pair works out its sales, at 90p each. (They can't sell more than they ordered.) They work out the cash in their box at the end of the day. Then you start another round.

◊ Sheet 147 contains a harder version where pupils are ordering hot dogs as well as ice creams.

'I tried this with Top and Bottom set. Both were hugely motivated and wanted to have several goes at it to improve their performances. We carried cash and supplies through week and worked out profits at end of week and had class winners.'

11 Parallel lines

p 54 **A** Looking for parallel lines	Introduction to what 'parallel' means
p 56 **B** Drawing parallel lines	
p 58 **C** Checking whether lines are parallel	

> **Essential**
>
> Sheets 152 and 153
> Plain paper
>
> **Practice booklet** pages 27 to 30

A Looking for parallel lines (p 54)

◊ This section introduces the term 'parallel' in a general way. Some pupils will have met the term before, so you could begin by asking pupils where they can see parallel lines in the classroom or elsewhere in the real world. You could then get a pupil to draw a pair of parallel lines on the board and ask the rest of the class to say what it is that makes the lines parallel. Clarify that parallel lines do not have to be the same length or lined up in some special way.

◊ The language of parallel lines needs to be developed and clarified carefully: 'a line parallel to *a*', 'they are parallel', 'a pair of parallel lines', 'a set of parallel lines'.

B Drawing parallel lines (p 56)

> Sheet 152, plain paper

◊ Although a method using the corner of a piece of paper is shown, pupils can use any suitable method for their patterns.

Many computer programs allow you to draw a line, copy it and drag it. Pupils can see that it remains parallel to the first line; they can also rotate it out of parallel. This may give pupils with a weak understanding of 'parallel' an important experience, and could be used to produce simple, attractive designs.

C Checking whether lines are parallel (p 58)

Sheet 153

People respond to the optical illusions on sheet 153 differently, so not all pupils will be equally surprised by them. When two lines turn out not to be parallel it is sometimes worth measuring to see how much further apart they are 'at one end' compared with the other.

A Looking for parallel lines (p 54)

A1 Line c

A2 (a) Line y

(b) Lines v and x are parallel.

A3 (a) True (b) False (c) True

A4 b and g are parallel;
c and i are parallel;
d and f are parallel.

A5 Line c with the pupil's explanation of how they decided.

B Drawing parallel lines (p 56)

B1 The pupil's drawings on sheet 152

C Checking whether lines are parallel (p 58)

C1 (a) The grey horizontal lines are parallel, though most people see the rows of tiles as tapering in alternate directions.

(b) The two black lines are not parallel! They are 7 mm further apart at the right-hand end.

C2 They are parallel.

What progress have you made? (p 58)

1 The pupil's parallel lines

2 They are not parallel.

Practice booklet

Section A (p 27)

1 Line e

2 (a) Line p (b) Line t
(c) Lines r, u, s are parallel.

3 m and h, g and k, c and e

4 (a) Line h (b) Lines b, d, i
(c) Line c (d) Line g

Section B (p 29)

1 The pupil's drawing

2 The pupil's triangle drawing

Section C (p 30)

1 Yes, despite appearances they are parallel.

2 No

3 d is parallel to a.

Anamorphs (p 59)

In addition to coordinate practice, this unit provides some important experience in visualisation.

Essential	Optional
Sheets 154 and 155	Coloured pencils
Practice booklet page 31	

◊ It may useful to allocate pictures and grids to pupils in such a way that each of the 16 possibilities is carried out in the class. The finished pictures make excellent wall displays, and pupils can then easily compare pictures on different grids.

◊ Pupils could draw their own pictures and then give a list of coordinates to a partner. It is important that using coordinates is not avoided.

◊ The skull is from 'The Ambassadors' (1533) by Hans Holbein and is in the National Gallery in London. The gallery sell an excellent CD-ROM which has a virtual reality demonstration of how the anamorph was designed to be seen by someone viewing it while walking up stairs towards it. Your art department may have it. Holbein is thought to have used a grid like the one on sheet 154 to paint the skull.

In the period following the second Jacobite rebellion (1745), portraits of Bonnie Prince Charlie were outlawed. Supporters often had anamorphic portraits made to avoid punishment.

If you have a scanner and suitable software a photo could be scanned in and distorted.

These are the pictures on a normal grid.

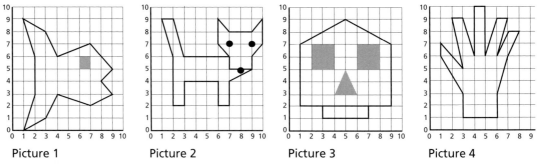

Picture 1   Picture 2   Picture 3   Picture 4

The message at the bottom of page 59 says (of course) SMP IS GREAT.

Practice booklet (p 31)

1 A (1, 9) B (7, 7)
   C (7, 9) D (10, 5)
   E (7, 1) F (7, 3)
   G (1, 1) H (2, 5)

2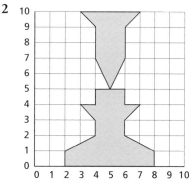

This is a well-known optical illusion where either a candlestick or two faces are seen.

Review 2 (p 60)

1 (a) The down rule is '– 2'.
        The missing number is 4.

   (b) The down rule is '+ 4'.
        The across rule is '– 5'.
        The missing number is 15.

   (c) The down rule is '+ 3'.
        The across rule is '– 2'.
        The missing number is 16.

2 (a)   (b)

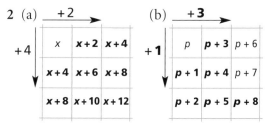

   (c)

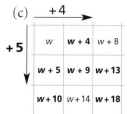

3 (a) $z + 9$ (b) $k + 9$ (c) $m + 8$

4 (a) 18   (b) 6.6 m   (c) 33 kg
   (d) 41 hours   (e) 154 cm

5 (a) Boys: 72 bpm   Girls: 78 bpm
   (b) Boys: 25 bpm   Girls: 18 bpm
   (c) Girls   (d) Boys

6 (a) b and e, f and l, g and h
   (b) a and j, d and i, c and k

7 The pupil's drawing

8 The pupil's drawing

Mixed questions 2 (practice booklet p 32)

1 (a)   (b)

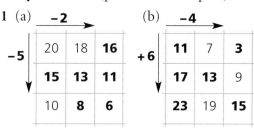

   (c)

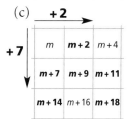

2 (a) (i) Girls 36 s   (ii) Boys 37 s
   (b) (i) Girls 50 s   (ii) Boys 36 s
   (c) (i) True   (ii) False   (iii) False

3 (a) d   (b) e

⑬ Practical problems

These problems use equipment so need to be done in class. Pupils need not do all the tasks: just doing some will tell you a lot about how well they can measure, estimate and apply number skills in problem solving.

One approach is to set up a 'circus' of tables; on each table is the equipment for one task and a label giving its name and page number (tasks with readily available equipment can be duplicated on more than one table). Pupils move round the circus following the instructions in the pupil's book. It is a good idea to have some 'exercise' work ready in case there is a log-jam as pupils go around the circus.

Alternatively, you can set up just one task (possibly in duplicate) and, while the rest of the class get on with written work, individuals or small groups come out in turn to do it.

'Weighty problems' is for a small group of pupils; the others can be done individually or in pairs.

Many pupils find weighing difficult, whether interpreting scale graduations on mechanical scales or coping with decimal places on digital balances. So it may be a good idea to add some straightforward weighing to the collection of tasks.

Weighty problems (p 62)

Scales or electronic balance
Two stones, say about 4 cm and 8 cm in diameter
A collection of familiar objects, including one with its weight clearly marked on it (for example a 500 g or 1 kg bag of sugar)

◊ In task 1 each group's estimates could be displayed on a dot plot (see 'Comparisons') and ideas of spread and over- or under-estimation discussed by the class. It is a good idea if everybody in the group checks the reading on the scales when the stones are weighed.

◊ After task 2 pupils could discuss ways of deciding which estimated order of weights was the best.

A related activity that goes well is for a pupil to hold the object of known weight (bag of sugar or whatever) in one hand and a different object in the other; the pupil estimates the weight of the other object then checks by weighing.

Beans (p 62)

> Two identical large sweet jars with lids, one empty and the other at least
> half filled with dried beans or pasta shapes with its lid taped down
> (butter beans are suitable, but not red kidney beans or other varieties
> that are poisonous when uncooked)
> About 100 extra beans or pasta shapes of the type in the jar
> An electronic balance or scales sensitive enough to weigh a few grams

◊ These are two approaches pupils have used to start solving the problem.

- Putting a layer of the extra beans into the empty jar and measuring the layer's height.
- Finding the weight of the beans in the jar by subtracting the weight of the empty jar.

In the second case, some go on to weigh a single bean. If so, ask them to check whether the beans all weigh the same. If they don't weigh the same, can pupils suggest a way to deal with this?

Cornflakes (p 63)

> A full box of cornflakes, with its price
> A cereal bowl
> An electronic balance or scales

◊ Some pupils may need a hint to work out the weight of the cornflakes without the bowl.

◊ You could extend the work to comparing the cost of different types of cereal or comparing the cost of an individual portion box with that of the same amount of cereal in a full-size box.

Getting better (p 63)

> A 5 ml spoon labelled 5 ml
> A container with a scale graduated in ml
> Three different sized medicine bottles (one less than 60 ml)
> distinguished by colour or labelling, but without their capacities marked
> Water, a tray and some paper towels

◊ A prepared answer sheet may help weaker pupils.

◊ Follow-up might include

- discussion of the appropriate level of accuracy
- estimation by pupils, perhaps as a homework assignment, of their daily fluid intake

Children's TV (p 64)

> *Radio Times*, or other paper with TV programme listings

◊ Some pupils may need to be reminded that there are 60 minutes in an hour.

◊ Pupils should be encouraged not to take easy options such as listing six half-hour programmes.

◊ Some teachers have asked their pupils to write a letter to their friend abroad to explain why they chose the programmes.

Follow-up for some pupils might include writing out a schedule for their tape, starting at 00:00.

Windfall (p 64)

> A shopping catalogue such as Argos or Littlewoods
> Optional: An order form for the pupils to fill in adds to the realism of the task

Be aware of pupils whose interest lies mainly in the contents of the catalogue, rather than the task in hand.

Angle dominoes (p 65)

This unit is intended as a revision of year 7 angle work. However, it is the first time the terms 'complementary' and 'supplementary' are used in the materials.

Essential

Sheet 156
Scissors

Practice booklet page 33

◊ Cutting out the pieces may take pupils some time. So you may prefer to photocopy the worksheet on to coloured card and save the pieces in sandwich bags to keep them for future use.

◊ The angles are often very close. Pupils should be encouraged to check that they add up to 90° or 180° as appropriate.

◊ Pupils could be asked what types of angle are needed to make a complementary/supplementary angle as revision of obtuse/acute.

Domino puzzle (p 65)

There are very many easily found solutions to this puzzle.

Practice booklet (p 33)

1 A and E, B and D, C and F

2 15° and 165° 55° and 125°
65° and 115° 75° and 105°
85° and 95°

Quadrilaterals

This unit helps pupils to recognise the names and explore the properties of the types of quadrilateral and to construct them accurately.

p 66	**A** Special quadrilaterals	Names and properties of quadrilaterals
p 69	**B** Quadrilaterals from triangles	Constructing quadrilaterals from triangles
p 70	**C** Accurate drawing	Drawing quadrilaterals from given dimensions

Essential	**Optional**
Scissors	Glue
Sheets 161 to 164	Sheet 164 on card
Coloured pencils	
Angle measurers	
Practice booklet pages 34 to 36	

A Special quadrilaterals (p 66)

> Sheets 161 to 163
> Coloured pencils
> Optional: Glue, scissors

In this section the names of the main quadrilateral types are introduced and their properties are explored.

◊ Many categories of quadrilaterals are subgroups of other categories. A square, for example, is a special type of rhombus. While this point is not emphasised at this early stage, pupils should not be told they are wrong, for example, if they identify a square as a rhombus, but that the term does not give a full description. This may arise in question A1.

◊ The mechanisms shown on page 67 may lead to discussion of why a parallelogram is used in such cases.

A4 In the answers, shapes are counted only once (for example, rhombuses are not included in parallelograms).

In the middle (p 68)

It is surprising that whatever quadrilateral you start with, the second shape is always some type of parallelogram. This is because opposite sides of the second shape are each parallel to and half of one of the diagonals of the original shape – though this observation is not expected of pupils.

B Quadrilaterals from triangles (p 69)

> Sheet 164, scissors

In this section pupils explore how different quadrilaterals can be made from different types of triangle. As well as reinforcing the names and properties of the quadrilaterals it also introduces the important idea that all quadrilaterals can be cut into two triangles.

◊ The triangles could either be copied on to card for pupils to draw round, or multiple paper copies could be made so that pupils are able to stick the results of their investigations into their books.

C Accurate drawing (p 70)

> Angle measurers

The emphasis in this section is on acquiring accurate construction skills.

A Special quadrilaterals (p 66)

A1–2 The pupil's drawings

A3 The pupil's shaded rectangles
There are 3 sizes of rectangle.

A4 The pupil's shaded quadrilaterals

(a) 2 sizes   (b) 3 sizes   (c) 2 sizes
(d) 2 sizes   (e) 6 sizes

In the middle (p 68)

(a) The original shape is a parallelogram.
The second shape is a parallelogram.
The third shape is a parallelogram.
The fourth shape is a parallelogram.

(b) The original shape is a kite.
The second shape is a rectangle.
The third shape is a rhombus.
The fourth shape is a rectangle.

(c) The original shape is a trapezium.
The second shape is a parallelogram.
The third shape is a parallelogram.
The fourth shape is a parallelogram.

(d) The original shape is a square.
The second shape is a square.
The third shape is a square.
The fourth shape is a square.

B Quadrilaterals from triangles (p 69)

B1 A is isosceles,   B is equilateral,
C is right-angled,   D is scalene.

B2 These quadrilaterals are possible.

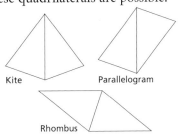

Kite   Parallelogram

Rhombus

B3 Only this rhombus is possible.

B4 These quadrilaterals are possible.

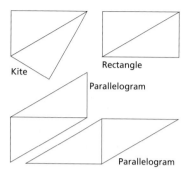

B5 (a) These quadrilaterals are possible.

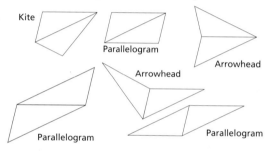

(b) The arrowheads could not be made.

B6 Only these are possible

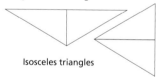

Isosceles triangles

B7 These are some possible answers.

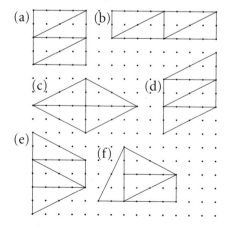

B8 (a) $\frac{1}{2}$ (b) $\frac{1}{4}$ (c) $\frac{1}{4}$ (d) $\frac{1}{3}$
(e) $\frac{2}{3}$ (f) $\frac{1}{6}$ (g) $\frac{1}{2}$

ℂ **Accurate drawing** (p 70)
The pupil's drawings

What progress have you made? (p 71)

1 Shape B   2 Shapes A and D

3 Kite, arrowhead
(and a trapezium sometimes has)

4 Square   5 The pupil's drawing

6 The pupil's drawing

Practice booklet

Section A (p 34)

1 (a) Parallelogram   (b) Kite
(c) Trapezium   (d) Trapezium
(e) Parallelogram (f) Rhombus

2 (a) BAFN, AKNM, HDIN or AHLI
(b) GAJN or HDIL
(c) AHNI
(d) BEIH or HCFI or ABNM or AFNK
(e) HDIA or HLIN
(f) BFIH or NHIM or BAIH or
AFIH or NKHI

3 (a) Parallelogram (with sketch)
(b) Kite (with sketch)

Section B (p 35)

1 (a) Both isosceles (b) Both scalene (c) Yes

2 (a) Either both right-angled or both scalene
(b) Either both isosceles or both scalene
(c) Both scalene

Section C (p 36)

1 The pupil's drawings and labels
(a) Kite   (b) Rhombus
(c) Parallelogram   (d) Trapezium

16 Is it an add?

In this unit pupils think about, discuss and choose appropriate operations and calculations for number problems. Although they may be able to solve contextual problems where the operation is clear, pupils often have difficulty choosing the right operation and calculation for a number problem.

p 72	**A** Add, subtract, multiply or divide?	Discussing the choice of the appropriate operation for a problem
p 76	**B** Video cassettes	Choosing an appropriate calculation
p 77	**C** Telling tales	Choosing calculations for written questions

Practice booklet pages 37 to 40

A Add, subtract, multiply or divide? (p 72)

Pupils discuss and select the correct operations for given problems.

◊ Pages 72 and 73 could be used for an introductory class discussion. First ask pupils, working in pairs or small groups, to decide, for each picture question on page 72, whether they would add, subtract, multiply or divide. They could try to explain why the operations they choose are correct.

You should try to bring out the features of each picture – in particular for multiplication

lots of identical things making a total

and division

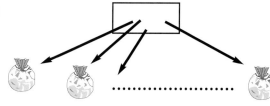

a total broken down into identical bits

'We used these ideas and worked through these pages together, which they all enjoyed, and they liked giving answers.'

Page 73 leads into the rest of the unit by asking which calculation from a set they would do to find the answer to each question. Ask them to try to explain why the calculation they have chosen for each question is correct.

◊ Encourage discussion about pages 74 and 75. In some cases there may be more than one correct answer, for example 20×3 and 3×20.

> 'It took a while for pupils to realise they did not have to work out the answers.'

$\mathbb{B}$ Video cassettes (p 76)

◊ Make it clear to the pupils that they have to write the calculations, not the answers (for example, '5×3' for B1, not '15').

$\mathbb{C}$ Telling tales (p 77)

C3 Pupils may find writing their own tales difficult. They could write these as questions like C1 and ask other pupils to do them.

$\mathbb{A}$ Add, subtract, multiply or divide? (p 72)

A1 8×20 or 20×8

A2 $80 \div 8$

A3 $80 \div 20$

A4 20×12

A5 12×8 or 8×12

A6 $12 + 8$ or $8 + 12$

A7 12×8 or 8×12

A8 $120 \div 12$

A9 $12 - 8$

A10 $20 - 8$

A11 30×12 or 12×30

A12 $30 - 12$

A13 $120 \div 30$

$\mathbb{B}$ Video cassettes (p 76)

B1 5×3 or 3×5

B2 $6 \div 4$

B3 $6 + 16$

B4 $16 - 14$

B5 4×14

B6 $11 - 8$

B7 $5 + 8$

B8 5×8

B9 $4 - 1.5$

B10 $15 \div 3$

$\mathbb{C}$ Telling tales (p 77)

C1 $20 \div 4$

C2 (a) $35 - 23$ (b) $96 \div 12$
    (c) $15 + 8$ (d) 18×5

C3 The pupil's tales

What progress have you made? (p 77)

1 (a) 6×12 (b) $150 \div 6$
   (c) $30 - 12$ (d) $150 + 30$

2 6×4

Practice booklet

Section B (p 37)

Soft drinks

1 12 − 5
2 6 + 12
3 3 × 3
4 2 + 3
5 12 × 4
6 5 − 3
7 30 ÷ 6
8 12 ÷ 2

Films

1 2 + 3
2 6 − 2
3 6 − 2
4 3 × 8
5 36 × 2
6 36 − 24
7 3 × 3
8 180 ÷ 36
9 8 ÷ 2
10 18 ÷ 6

Audio tapes

1 5 + 10
2 10 − 7
3 8 − 5
4 90 − 60
5 5 × 2
6 10 − 8
7 5 × 60
8 2 × 90
9 300 ÷ 60
10 15 ÷ 5

Section C (p 40)

1 (a) 12 × 6
   (b) 120 ÷ 6
   (c) 15 + 6
   (d) 24 − 5
2 (a) 40 × 7
   (b) 24 − 8
   (c) 420 ÷ 60
   (d) 15 + 17
   (e) 2000 ÷ 400
   (f) 64 × 25
3 The pupil's tales

⑰ Desk tidy (p 78)

This unit is one of a series of measurement activities requiring pupils to measure to a required degree of accuracy.

This unit requires pupils to measure to the nearest mm.

Essential	Optional
Rulers marked in cm and mm	A4 pieces of card Scissors

'I asked students to draw a scale plan first, they then constructed the tray as a series of small boxes set inside a larger box.
They drew nets at first on squared paper to avoid wasting card.'

Measurements can be recorded in millimetres and converted into decimals of a centimetre or vice versa.

◊ As an alternative pupils could make a tray for make-up, fishing tackle or hobby tools and could bring in the items for measuring or could measure them as homework.

A real challenge might be to design a tidy for the teacher's desk!

18 Frequency

This unit introduces frequency, frequency tables and bar charts. Both qualitative and discrete data are used.

The unit particularly emphasises the use of frequency when dealing with numeric data, as this can cause confusion for some pupils. Pupils are encouraged to write interpretations from data to enhance their report-writing abilities.

p 79	**A** Keeping order	Compiling data in frequency tables
p 81	**B** Names, words and letters	Reading information from frequency bar charts
p 83	**C** The mode	Finding the mode from a frequency bar chart
p 84	**D** How many times?	Describing frequency with discrete numeric data
p 86	**E** Cars and people	Using frequency to compare sets of data

Essential	**Optional**
Copy of a recent Saturday's football results from a newspaper Sheet 165	Tape of a recent Saturday's football results OHP transparency of football results Stopwatches and clipboards for car survey
Practice booklet pages 41 to 45	

A Keeping order (p 79)

'A good introduction but several in the group were inaccurate in counting and reluctant to use tallying.'

The essential objectives are that pupils learn how to put qualitative information (e.g. home win/away win/draw) into frequency tables.

At the end of this section they should be familiar with the term 'frequency'.

Newspaper cutting of a recent Saturday's football match; sheet 165
Optional: Tape recording of football results from radio/TV; transparency of newspaper cutting

◊ Give pupils the football results either as a photocopy or by playing a tape. A transparency of the results would be useful for working with the class. Usually the first two leagues are sufficient. They are asked to record whether the matches were home wins, away wins or draws. They will have met a tally chart in year 7. The idea of a frequency table can then be introduced.

◊ Questions A1 to A4 use sheet 165.
Pupils might be interested in finding out what some of the teams are called today.

B Names, words and letters (p 81)

This section introduces frequency bar charts.

How gud is yor spelin? (p 81)

◊ Ask pupils to write down the words 'isosceles' and 'parallelogram' on a scrap of paper. Use the collected scraps to compile a frequency table of spellings on the board. Ask questions such as
'How frequently was the word ... spelt?'
'What was the most frequent spelling?'
'What spelling had a frequency of ... ?'
A frequency bar chart could also be introduced at this stage.

B4 This is intended to be done without using a frequency table: there are too many letters to make use of a table sensible.

C The mode (p 83)

This short section for individual work simply introduces the mode as a term for the 'most frequent'.

◊ The term 'vowel' is avoided as some may argue about 'y'.

◊ Some pupils may need help with using the term 'modal' as an adjective.

D How many times? (p 84)

Some pupils become confused when they are dealing with the frequency of numerical data. This section tries to clarify any misunderstandings of this point.

◊ Give pupils some other statements about the ladybird chart and ask them to say whether each statement is true. Ask them to make some statements about the chart.

D2 To properly assess 'oktas' you would need to view the whole sky!

◊ If pupils need further practice in this area they could use the historical football data on sheet 165 to see if more goals were scored in matches in the earlier years.

E Cars and people (p 86)

This section consolidates all the ideas met in this unit and uses them to make comparisons with two similar sets of data. Pupils are encouraged to make statements by looking at frequency graphs as they will need to be able to do this later on for report writing in practical coursework.

◊ In the first activity pupils are asked to discuss how valid the statements are. They could be asked to make up some statements of their own and ask others in the group to check their validity

◊ Pupils could carry out a similar survey of their own and this section would then consist of a mini project. If direction is not sensible, surveys could be carried out a.m./p.m. or in rush hour and not.

A Keeping order (p 80)

A1 (a) 10   (b) 1

	A1(c) 1900	A2(a) 1930	A2(b) 1950
Home win	10	26	15
Away win	1	10	5
Draw	3	8	13

A3 (a) Home win   (b) Home win

(c) Home win   (d) Own result

A4 (a) 3   (b) 8   (c) Draw

B Names, words and letters (p 81)

B1 (a) 10   (b) 14

(c) 'fourty'   (d) 49

B2 (a) 16   (b) 'succes'   (c) 23

B3 (a) Willy

(b) The frequency of that choice

(c) 12   (d) 1

B4 (a) Wales (11!)   (b) Wales (13)

C The mode (p 83)

C1 (a) e   (b) a   (c) u

(d) e, a, o, i, u

C2 (a) e

(b) (i) True   (ii) False

(iii) True

D How many times? (p 84)

D1 (a) (i) 8   (ii) 3   (iii) 1

(b) 3 spots   (c) 45   (d) 10

D2 (a)

Cloud cover	No. of days
0	4
1	5
2	2
3	3
4	1
5	1
6	3
7	6
8	5

(b)

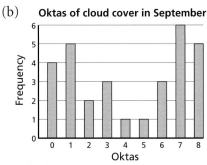

Oktas of cloud cover in September

(c) 7 oktas

Cars and people (p 87)

E1 (a) 14   (b) 68

E2 (a)
People	Frequency
1	28
2	15
3	6
4	1

(b)

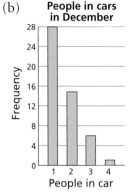

People in cars in December

(c) The modal number of people is 1. There were never more than 4 people in a car.

E3 Statements such as:
'The mode for July is 3, and for December it is 1.'
'In July, the highest number of people in a car was 5, but in December it was 4.'
'More families on holiday were around in July, but in December it was local and business people.'

What progress have you made? (p 88)

1
Pieces of fruit	Frequency
0	5
1	10
2	8
3	4
4	2
5	1

2 (a) 5   (b) 7   (c) 30

3

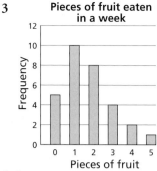

Pieces of fruit eaten in a week

4 1

5 Two statements such as:
'The mode in year 10 was 0, 1 in year 8.'
'3 year 10 pupils ate 5 pieces but only 1 year 8.'
'Year 8s eat more fruit on the whole.'

Practice booklet

Section A (p 41)

1 (a)
Colour	Frequency
Brown	18
Blonde	5
Black	3
Auburn	1

(b) Brown   (c) 5   (d) Auburn

2 (a) Brown   (b) 9

Section B (p 42)

1 (a) Black   (b) 20   (c) Red

2 (a)
Zone	Frequency
Gold	2
Red	5
Blue	2
Black	1
(White	0)

(b) Red

(c)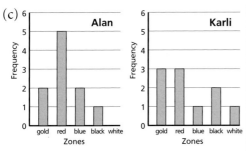

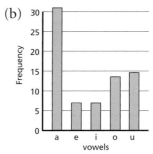

(d) Alan gets more in higher zones, but Karli gets more golds. It is difficult to say who did best.

(e) Alan 66   Karli 60

Section C (p 43)

1 (a) Brimstone   (b) Cabbage White
(c) 6

2 (a)

Letter	Frequency
a	31
e	7
i	7
o	13
u	14

(b)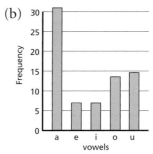

(c) a
(The poem is from 'Where the Red Lehua Grows' by Jane Comstock.)

Section D (p 44)

1 (a) 4   (b) 4   (c) 25

2 (a) 4   (b) 6   (c) 0

Section E (p 45)

1 (a) 2   (b) 6   (c) 1 visit
(d) 0 visits   (e) 5
(f) (i) 19   (ii) 16

2 (a) True   (b) False
(c) True   (d) True

Amazing but true! (p 89)

This unit revises work on coordinates. Later units will introduce negative coordinates.

The maze shown in the picture, based on a castle, was designed and made by Adrian Fisher, a professional maze designer.

Essential

Squared paper

Practice booklet page 46

◊ You might wish to use the maze shown on page 89 with the whole class before pupils design their own mazes.

◊ An early check that all pupils are giving *x*-coordinates first is essential. Deciding whether to go left or right is a good visualisation skill. As an alternative, compass directions could be substituted.

Start at (0, 1)	Go straight across	At (3, 5)	Go left
At (9, 1)	Go left	At (3, 7)	Go right
At (9, 7)	Go left	At (5, 7)	Go right
At (8, 7)	Go left	At (5, 6)	Go left
At (8, 2)	Go right	At (6, 6)	Go right
At (2, 2)	Go right	At (6, 5)	Go left
At (2, 5)	Go right	At (7, 5)	Go right
At (3, 5)	Go right	At (7, 3)	Go left
At (3, 3)	Go left	At (8, 3)	Go left
At (7, 3)	Go left	At (8, 8)	Go left
At (7, 5)	Go left	At (2, 8)	Go left
At (6, 5)	Go right	At (2, 7)	Go right
At (6, 6)	Go left	At (1, 7)	Go right
At (5, 6)	Go right	At (1, 9)	Go right
At (5, 7)	Go left	At (9, 9)	Go right
At (3, 7)	Go right	At (9, 1)	Go left to exit!
At (3, 9)	Go right, to the exit		

Practice booklet (p 46)

1 Start at (0, 1) — Go straight across
At (4, 1) — Go left
At (4, 2) — Go left
At (2, 2) — Go right
At (2, 5) — Go right

2

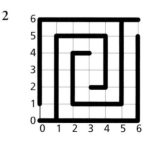

Review 3 (p 90)

1 (a) Shape B

(b) Shapes A and G

(c) Shape D

(d) Shapes B and F

(e) Shapes C and D

(f) Shape E

(g) Right-angled (isosceles) triangles

2 (a) The pupil's drawing of shapes

(b) (i) Parallelogram

(ii) Kite

3 (a) 90×3 (b) $80 \div 4$

(c) 12×6 (d) $12 - 4$

4 (a)

Number of brothers and sisters	Frequency
0	3
1	5
2	8
3	4
4	1
5	0
6	1

(b) 2

(c)

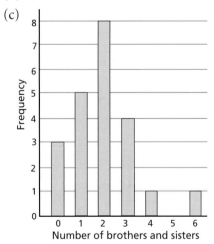

5 (a) 24   (b) 1

(c) 3

(d) At Waveney the mode is higher but they have a smaller range.

Mixed questions 3 (practice booklet p 47)

1 (a) A   Trapezium

B   Kite

C   Rhombus (parallelogram)

D   Arrowhead

E   Rectangle

F   Parallelogram

(b) A and F

(c) B and D

2 $60 \div 4$

3 (a)

Injuries	Frequency
0	3
1	6
2	7
3	9
4	4
5	1

(b) 4

(c) 3

(d)

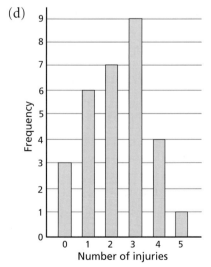

20 Photo display (p 92)

For weaker pupils, this unit will give practice in measuring to the nearest millimetre or 0.1 cm. Those who are more confident will move on to work out for themselves how to find the widths of the margins when a photo is to be placed in the centre of a card.

Essential	Optional
Photographs of various sizes	OHP
Card	Pupils' own photographs that they are willing to stick on card

◊ Start by getting pupils to measure the height and width of some photos to the nearest millimetre, writing their measurements both in millimetres and as a decimal of a centimetre. Take the opportunity to see if any pupils still have difficulty with measuring of this kind.

As well as photos you could use postcards or collector's cards (football cards or whatever is in vogue).

Now take a photo and place it on a larger rectangle. This can be done on an OHP (where the photo will show as a dark rectangle) or as a desktop demonstration to a smaller group of pupils. Ask pupils to tell you when they think the photo is in the middle of the 'frame'; then ask them to measure the border each side and top and bottom to see if they were right. By repeating this with other sized photos, build up to asking pupils to suggest methods for calculating the border widths.

◊ If pupils find it difficult to work out the border size, an initial activity might be to get them to measure one picture, and then to work out how large a piece of card needs to be if there is to be, for example, a 2 cm border all around the picture.

◊ If pupils have mastered how to work out border sizes, then they may be ready to mount pictures of their own on to pieces of card. They should calculate exactly where the picture needs to be placed, and draw its outline lightly in pencil. They can then use the outline to check their calculation.

Note that when using their own pictures, calculations of the type $2.7 \div 2$ will arise, where the result has two decimal places. You will need to discuss with pupils how to deal with such results.

1 (a) 11.7 cm   (b) 11.2 cm

2 (a) Top and bottom 7.6 cm
        Sides 10.1 cm

(b) Top and bottom 2.8 cm
       Sides 2.6 cm

(c) Top and bottom 4.9 cm
       Sides 3.9 cm

21 Fractions 1

This unit revises fractions of shapes, first of the form $\frac{1}{n}$ then $\frac{m}{n}$.
Emphasis is placed on equality of parts.

T	p 93 **A** Equal parts?	Recognising fractions of shapes of the form $\frac{1}{n}$
T	p 94 **B** More than one part	Recognising fractions of shapes of the form $\frac{m}{n}$
	p 96 **C** Coloured tiles	Finding and drawing fractions of shapes

Essential	**Optional**
Squared paper	Sheet 166
Practice booklet pages 48 and 49	

A Equal parts? (p 93)

T

The purpose of the introduction is to establish that when identifying fractions of shapes the parts involved must be equal. The fractions are all of the form $\frac{1}{n}$.

The whole of section A might be done orally, with pupils being encouraged to explain their conclusions. Pupils may base their explanations either on splitting each shape into (possibly) equal parts, or on counting squares.

B More than one part (p 94)

T

Pupils move on to consider fractions of shapes of the form $\frac{m}{n}$.

Again the whole of the section might be done orally.

C Coloured tiles (p 96)

'We established that if all the lines to give "equal parts" were not there, then they should add them. This worked well.'

Squared paper
Optional: Sheet 166 (for question C1)

The fact that some fractions are equivalent may come up naturally. It is not essential for the section, but you may welcome the opportunity to discuss equivalent fractions if it arises.

Ⓐ **Equal parts?** (p 93)

A1 (a) No (b) Yes (c) Yes

A2 (a) No (b) Yes (c) Yes (d) No

A3 (a) Yes (b) No (c) Yes (d) Yes

A4 (a) Yes (b) Yes (c) No

A5 (a) Yes (b) No (c) No

Ⓑ **More than one part** (p 95)

B1 (a) Yes (b) No (c) No

B2 (a) Yes (b) No (c) Yes (d) No

B3 (a) No (b) Yes (c) No (d) Yes
(e) No

B4 (a) Yes (b) No (c) No

B5 (a) No (b) No (c) Yes

Ⓒ **Coloured tiles** (p 96)

C1 The coloured fractions are
(a) $\frac{1}{2}$ (b) $\frac{5}{8}$ (c) $\frac{2}{3}$ (d) $\frac{1}{4}$
(e) $\frac{1}{2}$ (f) $\frac{3}{5}$ (g) $\frac{4}{10}, \frac{2}{5}$ (h) $\frac{3}{8}$
(i) $\frac{3}{8}$ (j) $\frac{3}{9}, \frac{1}{3}$ (k) $\frac{1}{10}$ (l) $\frac{4}{10}, \frac{2}{5}$
(m) $\frac{4}{8}, \frac{1}{2}$ (n) $\frac{3}{8}$ (o) $\frac{6}{8}, \frac{3}{4}$ (p) $\frac{3}{8}$
(q) $\frac{1}{16}$ (r) $\frac{5}{16}$ (s) $\frac{6}{16}, \frac{3}{8}$ (t) $\frac{4}{16}, \frac{1}{4}$
For each tile,
white fraction = 1 − coloured fraction

C2 The pupil's diagrams

C3 The pupil's diagrams

C4 The pupil's tiles

C5 $\frac{1}{4}$

C6 The pupil's design

What progress have you made? (p 97)

1 (a) No (b) Yes

2 (a) $\frac{3}{8}$ (b) $\frac{5}{8}$

Practice booklet

Sections A and B (p 48)

1 A, B and D have $\frac{1}{4}$ shaded.

2 A, C and E have $\frac{2}{3}$ shaded.

Section C (p 49)

1 The grey fractions are
(a) $\frac{1}{3}$ (b) $\frac{3}{4}$ (c) $\frac{2}{5}$ (d) $\frac{3}{8}$
(e) $\frac{7}{10}$ (f) $\frac{5}{8}$ (g) $\frac{4}{9}$ (h) $\frac{8}{12}, \frac{2}{3}$
(i) $\frac{7}{18}$ (j) $\frac{7}{16}$ (k) $\frac{7}{8}$ (l) $\frac{3}{10}$

For each tile,
white fraction = 1 − grey fraction

2 The pupil's drawings

3 The pupil's drawings

22 Enlargement

Essential	Optional
Centimetre squared paper Sheets 167 to 169	OHP and squared grid
Practice booklet pages 50 and 51	

A Spotting enlargements (p 98)

T

'This led to a really good discussion – and made these important points about enlargement.'

Pupils should not need to measure any of the photographs to detect the fakes, all of which involve only distortions of the fish. Pupils should be encouraged to explain why they are not enlargements, either by using reference points on Bert or by comparing proportions.

B, D and G are enlargements, the rest are not.

The question on page 100 could be done by pupils working in small groups, and their answers discussed. Pupils' explanations will need to be more numerical. A common error is to regard D as a correct enlargement as both horizontal and vertical bars have increased by one square.

B, E and H are enlargements, the rest are not.

B Enlarging shapes (p 101)

Centimetre squared paper, sheets 167 to 169
Optional: OHP and squared grid

T

◊ You might wish to start your discussion of how to enlarge shapes with a simpler example (with no sloping lines) than that on page 101.

◊ Some pupils find it difficult to draw diagonals, particularly when they are not at 45°, and may need some extra practice.

They may find an explanation like this helpful:

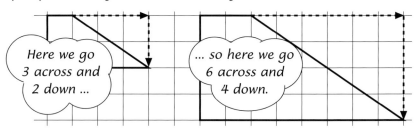

Here we go 3 across and 2 down ...

... so here we go 6 across and 4 down.

'We did our own designs for display. The pupils were very creative.'

◊ The 'Enlarging designs' activity on page 102 is useful consolidation work and is worth spending some time on.

ⓑ Enlarging shapes (p 101)

B1 The pupil's enlargements

B2 The pupil's enlargements

B3 The pupil's enlargements

B4 T scale factor 4   U scale factor 5
V scale factor 3

B5 The pupil's enlargement of S,
scale factor 2

B6 (a) Q is an enlargement.

    (b) Scale factor 2

What progress have you made? (p 103)

1 Shape B

2 Scale factor 2

3 (a) The pupil's enlargement,
scale factor 2

    (b) The pupil's enlargement,
scale factor 3

Practice booklet

Section A (p 50)

  1 C and H

  2 E and F

Section B (p 51)

  1, 2 and **3** The pupil's drawings

㉓ Calculate in order

Pupils are introduced to the priority rules for multiplication, division, addition and subtraction.

They use a calculator to evaluate expressions with and without brackets.

T	p 104 **A** In order	Establishing the priority rules for calculations
T	p 106 **B** Brackets	Using brackets to control the order of operations
T	p 108 **C** All keyed up	Using calculators to evaluate expressions with and without brackets

Essential	Optional
Calculators that use priority rules Sheets 178 and 180	Sheet 181
Practice booklet pages 52 and 53	

A **In order** (p 104)

T

The objective of the teacher-led discussion is to arrive at the priority rules for dealing with multiplication, division, addition and subtraction.

A game consolidates the use of these rules.

◊ The result of each calculation A to K is either 2 or 8. The calculations can be used to introduce the priority rules in a variety of ways.

◊ Almost all calculators now use the priority rules and it is important that pupils are not using a calculator which works from left to right. Ask pupils to use their calculators to find the result of each calculation A to K. Point out that they must not press the '=' sign until the end of the calculation. Pupils could try to produce a description of the rules they think the calculators are using. A brief statement of the rules could be:

• They multiply or divide before they add or subtract.

• Otherwise, they work from left to right.

Alternatively, you could begin without calculators, asking pupils to find any calculations with result 2 or 8. Most pupils will consistently work from left to right and will not include B, C, E, I or J. Ask if they can think of a way of getting the result 2 or 8 for each of these. Some may suggest using brackets to show which part of the calculation needs to be done first. Point out that brackets are not needed if we use the rules stated above.

◊ Once pupils understand the priority rules, point out that these are used as standard in numerical calculations.

◊ Some calculators use the symbols * and / for × and ÷ respectively and this may need pointing out. Spreadsheets use this same notation.

◊ Pupils could use a calculator to check their results for A1 to A4.

*A4 Pupils could make up their own puzzles like this.

Operation 3062 (p 105)

This game consolidates the priority rules for multiplication, division, addition and subtraction.

> Sheet 178 (one board each so one copy for each pair)
> Sheet 180 (one copy each)
> Optional: sheet 181

◊ Calculators should not be used for the game.

◊ A full set of possible results for *Operation 3062* is

$30 + 6 + 2 = 38$	$30 + 6 - 2 = 34$
$30 + 6 \times 2 = 42$	$30 + 6 \div 2 = 33$
$30 - 6 + 2 = 26$	$30 - 6 - 2 = 22$
$30 - 6 \times 2 = 18$	$30 - 6 \div 2 = 27$
$30 \times 6 + 2 = 182$	$30 \times 6 - 2 = 178$
$30 \times 6 \times 2 = 360$	$30 \times 6 \div 2 = 90$
$30 \div 6 + 2 = 7$	$30 \div 6 - 2 = 3$
$30 \div 6 \times 2 = 10$	$30 \div 6 \div 2 = 2.5$

Alternative versions

• Play the game so the person with the **lowest** score wins.

• Each pupil has one board (sheet 178), eight operation cards (half of the cards on sheet 180; two each of ×, ÷, + and –) and one set of result cards (sheet 181). Individuals try to find a way to make each number on the result cards.

• A harder version of this game is described on sheet 179.

B Brackets (p 106)

Pupils should do the questions without a calculator but could use a scientific calculator to check their results.

C **All keyed up** (p 108)

Calculators that use priority rules

Pupils use a calculator to evaluate expressions with and without brackets.

◊ Discuss how pupils could use their calculators to evaluate $(6 + 13) \times 7$ and $2 + 3 \times 17$. The working on the page shows intermediate steps and you may wish to encourage pupils to set out their work like this.

◊ Pupils whose calculators have brackets could make use of these keys for C1 and you need to decide if you want to encourage this at this stage. They could also just key in the calculations in C2 and avoid showing any intermediate steps in their working. Again, you need to decide if you want to encourage this at this stage.

A **In order** (p 104)

A1 P, Q, T and U

A2 A and E (3)   B and C (5)   D and J (15)
    F and H (10)   G and I (7)

A3 (a) 4   (b) 7   (c) 10   (d) 2
    (e) 4   (f) 7   (g) 8   (h) 12
    (i) 7   (j) 8   (k) 9   (l) 12
    (m) 11   (n) 2   (o) 0

*****A4** (a) $8 + 6 - 2$ (b) $8 \times 6 - 2$
    (c) $8 + 6 \times 2$ (d) $8 + 6 \div 2$

B **Brackets** (p 106)

B1 (a) 8   (b) 3   (c) 12   (d) 15
    (e) 2   (f) 2   (g) 2   (h) 6
    (i) 30   (j) 7   (k) 4   (l) 36
    (m) 3   (n) 4   (o) 2   (p) 5

B2 A and J (9)   B and D (4)   C and F (11)
    E and I (2)   G and H (6)

B3 (a) 12   (b) 14   (c) 20   (d) 14
    (e) 21   (f) 1   (g) 5   (h) 7
    (i) 5   (j) 11   (k) 10   (l) 0
    (m) 5   (n) 3   (o) 3   (p) 7
    (q) 4   (r) 1

B4 (a) 3   (b) 3   (c) 4   (d) 6
    (e) 6   (f) 3   (g) 4   (h) 4
    (i) 6   (j) 3   (k) 6   (l) 4

*****B5** (a) 3   (b) 6   (c) 3   (d) 11
    (e) 12   (f) 6   (g) 15   (h) 3

C **All keyed up** (p 108)

C1 (a) 102   (b) 261   (c) 30
    (d) 9   (e) 202   (f) 8
    (g) 6   (h) 336   (i) 11

C2 (a) 182   (b) 391   (c) 80
    (d) 86   (e) 114   (f) 554
    (g) 2   (h) 138   (i) 42

C3 (a) 624   (b) 68   (c) 51
    (d) 9   (e) 6   (f) 302
    (g) 30   (h) 60   (i) 1904

C4 (a) 24   (b) 204.7   (c) 160
    (d) 36   (e) 6.7   (f) 42
    (g) 9.6   (h) 51.5   (i) 0.7
    (j) 21.5   (k) 9.5   (l) 25
    (m) 73.8   (n) 2.1   (o) 134.1

What progress have you made? (p 109)

1 A and E (15)   B and D (7)   C and G (3)
  F and H (6)

2 (a) 14   (b) 16   (c) 6
  (d) 9   (e) 1   (f) 1

3 (a) 16   (b) 2   (c) 1
  (d) 18   (e) 4   (f) 2

4 (a) 20   (b) 17   (c) 2
  (d) 6   (e) 2   (f) 11

5 (a) 15   (b) 33   (c) 5
  (d) 8   (e) 227   (f) 117
  (g) 22.5 (h) 103   (i) 41
  (j) 125

Practice booklet

Section A (p 52)

1 (a) 18 (b) 18 (c) 24 (d) 24
  (e) 24 (f) 18 (g) 24 (h) 18

2 (a) 6   (b) 11 (c) 17 (d) 6
  (e) 11 (f) 7   (g) 13 (h) 1

3 E – all the rest equal 8.

Section B (p 52)

1 (a) 12 (b) 20 (c) 12 (d) 12
  (e) 20 (f) 20 (g) 20 (h) 20

2 (a) 18 (b) 15 (c) 4   (d) 3
  (e) 15 (f) 2   (g) 4   (h) 5

3 (a) 5   (b) 8   (c) 2
  (d) 2   (e) 8   (f) 2

4 A and F (14)   B and H (17)
  C and G (5)   D and E (11)

Section C (p 53)

1 (a) 85   (b) 322   (c) 119
  (d) 176   (e) 21   (f) 172
  (g) 54   (h) 17   (i) 406

2 (a) 42   (b) 248.8 (c) 275
  (d) 85   (e) 5.9   (f) 80
  (g) 8.5   (h) 32   (i) 19.5

3 (a) Correct   (b) Wrong, 48
  (c) Wrong, 19   (d) Wrong, 32.2

Graphs and charts

Pupils interpret graphs and charts from real-life sources and draw their own frequency bar charts and line graphs.

T	p 110 **A** Children's income	Interpreting charts
	p 111 **B** Shut up!	Interpreting single line graphs
T	p 112 **C** Off the record	Interpreting more complex line graphs
	p 113 **D** Equal shares	Interpreting split bar charts
T	p 114 **E** Drawing graphs and charts	

Essential	**Optional**
Sheets 182 to 185	Wall's Monitor
	OHP transparency ruled as pupils' graph paper
	Sheet 186
Practice booklet pages 54 to 57	

A Children's income (p 110)

> Optional: Wall's Monitor (Available from Birds Eye Wall's Limited, Station Avenue, Walton-on-Thames, Surrey KT12 1NT; tel. 01932 263000. This was free at the time of writing.)

T

'Very good, relevant and they all have an opinion.'

'Wall's info led to considerable useful discussion and caught the pupils' interest.'

The chart shows the total weekly income of children, which is the sum of pocket money, hand-outs from other relatives and any earnings. Before you begin you may want to collect data from the class to compare with the data given in the chart. This can be a sensitive issue so try to avoid a show of hands: getting pupils to write their income on a piece of paper that you collect can alleviate problems.

The data from Wall's Monitor could be used when drawing graphs and charts later in the unit. Wall's Monitor is an annual pocket money survey and is based on data provided by Gallup. It provides trend data on children's income. It contains a host of data.

B Shut up! (p 111)

Comparing the information on the two graphs in this section can lead to a useful class discussion.

C Off the record (p 112)

T

You may need to explain what an LP or a single is.

'This worked well as a group activity.'

D Equal shares (p 113)

E Drawing graphs and charts (p 114)

Sheets 182 to 185
Optional: OHP transparency as pupils' graph paper, sheet 186

T

◊ You can use the data on the top of page 114 to introduce drawing a frequency bar chart. You may wish to discuss suitable intervals for grouping, but pupils are not expected to choose their own groupings in this unit (intervals of 0–9, 10–19 etc. will be fine). Note that this may be the first time that pupils have met the idea of grouping data.

If you use an OHP, you will find it useful to have a transparency ruled just like the graph paper the pupils use. Otherwise, a grid on the board is essential.

You will need to point out how to draw suitable bars on the chart.
Either label each bar '0–9', '10–19', leaving gaps between bars to emphasise the discreteness, or mark off a normal continuous scale.
If a data item falls on a boundary between class intervals, it is normally included in the upper interval.
No distinction has been made in this unit between representing discrete and continuous data.

◊ Only a few data sets are given in the pupils' book, as the work is more motivating if you use data relevant to the class.
You could use their own scores from a recent test, heights, reaction times or any data which can be grouped into convenient intervals.

◊ Sheets 182 and 184 provide follow-up examples which can be used individually or worked through with the teacher.
Sheets 183 and 185 are blanks for your own data.

◊ This is a tally chart for the data at the top of page 114.

Score	Tally	Frequency
0–9	/	1
10–19	///	3
20–29	////	4
30–39	ЖЖ ЖЖ /	11
40–49	ЖЖ	5
50–59	ЖЖ /	6
60–69	////	4
70–79	ЖЖ	5
80–89	/	1
	Total	40

E1–3 Sheet 186 may be useful for questions E1 to E3 if pupils find tallying difficult. One pupil can read out each data item and cross it out on the sheet, while another pupil makes a corresponding tally mark in the table.

E4, 5 These questions are best done on squared paper. If pupils find drawing the scales difficult, you might draw them yourself on squared paper and copy your sheet for pupils to use.

Ⓐ **Children's income** (p 110)

A1 (a) 605p   (b) Scotland
(c) North West
(d) Difference = 605p – 397p = 208p

A2 (a) 50%   (b) 25% to 30%

A3 (a) Sweets   (b) Sweets
(c) Cigarettes

A4 4629 pupils (written at bottom)

A5 The pupil's answers

Ⓑ **Shut up!** (p 111)

B1 (a) 1990
(b) (i) 1987 and 1988 (ii) About 60 000
(c) 1985 and 1986 (d) 1989 and 1990

B2 12 000 (roughly)

B3 For example, 'Between 1983 and 1985 complaints rose steadily. Then they dropped between 1985 and 1986. They rose in the next year and stayed the same between 1987 and 1988. They rose again between 1988 and 1989 and then rose very steeply up to 1990.'

B4 The scale is not continuous from zero.

B5 Between 1985 and 1986

B6 1986

Ⓒ **Off the record** (p 112)

C1 (a) The sales of CDs started in 1983 and went up every year.
(b) Sales of cassettes rose to just over 80 million in 1989, then dropped.
(c) 1989   (d) 1986
(e) About 80 million

C2 The pupil's questions.

Ⅾ **Equal shares** (p 113)

D1 (a) Repairs   (b) Washing and ironing

D2 Washing and ironing – it has the lowest percentage of 'mainly men' and 'shared equally' put together.

D3 Washing and ironing, cleaning house and making evening meal

D4 Disciplining children

D5 Washing dishes, paying bills and perhaps disciplining children all have roughly equal percentages of 'mainly men' and 'mainly women'.

Ⅰ **Drawing graphs and charts** (p 114)

E1

Age group	Tally	Frequency
0–9	/	1
10–19	//	2
20–29	⊬卅 /	6
30–39	////	4
40–49	////	4
50–59	⊬卅	5
60–69	⊬卅	5
70–79	////	4
80–89	/	1
	Total	32

The pupil's bar chart

E2

No. of cars	Tally	Frequency
50–59	⊬卅 /	6
60–69	⊬卅	5
70–79	////	4
80–89	⊬卅 ⊬卅	10
90–99	⊬卅 ///	8
100–109	////	4
110–119	///	3
	Total	40

The pupil's bar chart

E3

Reaction time	Tally	Frequency
10–12	⊬卅	5
13–15	⊬卅 //	7
16–18	⊬卅 ⊬卅 //	12
19–21	////	4
22–24	/	1
25–27	/	1
	Total	30

The pupil's bar chart

E4, 5 The pupil's graphs

Sheet 182

Age group	Tally	Frequency
0–9	///	3
10–19	⊬卅 /	6
20–29	⊬卅 //	7
30–39	⊬卅 /	6
40–49	⊬卅 /	6
50–59	///	3
60–69	⊬卅 /	6
70–79	//	2
80–89	/	1
	Total	40

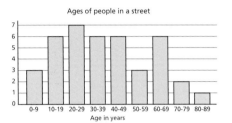

Ages of people in a street

Sheet 184

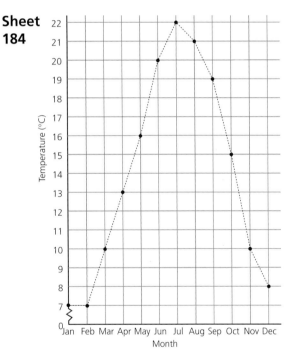

What progress have you made? (p 115)

1 (a) 13   (b) 8

2 (a) 38°C

   (b) It was 37°C at 3 p.m., then rose to 40°C at 6 p.m., dropping gradually to 38°C at 9 p.m. with a slight rise in between.

3

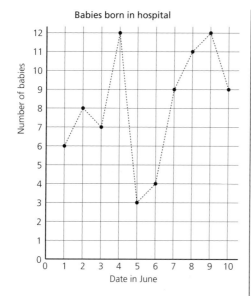

Babies born in hospital

4

Age group	Tally	Frequency
0–9	///	3
10–19	////	4
20–29	⧸⧸⧸⧸ /	6
30–39	///	3
40–49	//	2
50–59	/	1
60–69	////	4
70–79	////	4
	Total	27

5

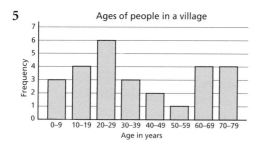

Ages of people in a village

Practice booklet

Section A (p 54)

1 (a) Bus   (b) Car   (c) 30%

(d) 8%   (e) Car   (f) Bike

(g) More than year 7

(h) More than year 7

Section C (p 55)

1 (a) London, 4°C   (b) Sydney, 22°C

(c) The temperature went down until the 7th and 8th; then went up until the 11th, and then went down again.

(d) The temperature went up until the 5th and 6th; then went down until the 10th and 11th; then went up again.

(e) 10°C   (f) 14°C

(g) On the 4th, 5th, 6th, 7th, 8th, 13th, 14th and 15th

Section D (p 56)

1 (a) The UK   (b) Greece

(c) Portugal and Greece

(d) The Netherlands

(e) Germany

(f) Spain, Portugal, Italy and Germany

(g) Spain, Portugal, Italy, Ireland, Greece and Germany

Section E (p 57)

1 (a)

Weight	Tally	Frequency
0–9	/	1
10–19	⧸⧸⧸⧸ /	6
20–29	⧸⧸⧸⧸	5
30–39	⧸⧸⧸⧸ ///	8
40–49	⧸⧸⧸⧸ /	6
50–59	////	4
	Total	30

(b) The pupil's bar chart

(c) The most common weight of fish caught in this pond is between **30** and **39** grams.

2 (a) The pupil's graph

(b) The rainfall is very low from December to April, then increases to June, drops a little in July, and goes up in August to October; then it drops once more.

㉕ Fractions 2

This unit revises finding fractions of numbers, starting with a problem-solving activity for the whole class.

T	p 117 **A** Chocolate	Problem-solving activity
T	p 117 **B** Simple fractions of numbers	Working out fractions such as $\frac{1}{2}$ of 1972
T	p 119 **C** Other fractions of numbers	Working out fractions such as $\frac{2}{3}$ of 792

Essential	**Optional**
6 bars or blocks of something, which can be divided up and shared out equally	Bars of chocolate
Practice booklet pages 58 to 61	

𝔸 **Chocolate** (p 117)

T

> 'I thought this would be chaos and pupils would gain little from it. In fact, lots of useful discussion was generated! Had I remembered to buy the chocolate it would have been even better. (We used sheets of paper to represent chocolate.)'

This is a problem-solving activity in which pupils can apply their understanding of fractions. Pupils decide which table to sit at in the hope of getting the most chocolate.

The activity provides opportunities for

- introducing pupils to the idea that a fraction like $\frac{2}{3}$ of a bar can mean '2 bars divided between 3 pupils'
- introducing methods of finding fractions of numbers such as $\frac{2}{3}$ of 24

6 bars or blocks of something, which can be divided up and shared out equally
Optional: Bars of chocolate

Getting started

◊ It is best to start with a fairly simple situation. For example:

> 'I split the class into three teams. Each team in turn sent one person to sit at the tables. The aim was for one team to get the most chocolate.'

- Distribute 3 chocolate bars on 2 tables as shown.
- Choose a group of, say, 8 pupils to take part.
- Tell them that you will ask them one by one to choose a table to sit at (they cannot change their minds later).
- Once they are all seated, they will get an equal share of the chocolate on their table.

Different groups of pupils can be chosen to give all pupils a chance to take part.

Solving the problem

◊ As pupils choose where to sit, involve the whole class and ask questions such as:
- Where do you think the next person should sit? How would you advise them?
- What fraction of a chocolate bar would each person get at this table if no one else sits here? (Emphasise that 2 bars divided between 3 people means that each person gets $\frac{2}{3}$ of a bar.)

◊ Once the last pupil has chosen, ask pupils to decide which pupils get the most chocolate and to justify their decision. Pupils could consider this in small groups and then explain how they decided to the whole class.

Variations
- Change the number of pupils who are to sit at the tables.
- Change the number of tables and/or the numbers of bars on each table. For example, use three tables – one with 1 bar, another with 2 and the third with 3.
- Once a group is seated at the tables, and who gets most is decided, ask if a different seating arrangement might make the division of the bars fairer.

> 'In the second lesson, 8 pupils simultaneously sat where they thought they'd get most chocolate. I gave them a chance to move. The rest of the class was the audience. This generated considerable discussion.'

Follow-up

The activity works well when pupils cannot see how many squares are in each chocolate bar. However, repeating the activity when pupils **can** see how many squares are in a bar will provide a useful lead-in to the rest of the unit. Choose (or draw) bars of chocolate that have a number of squares with plenty of factors, such as 24.

B Simple fractions of numbers (p 117)

T

You could give pupils some oral fractions questions, such as $\frac{1}{2}$ of 42, $\frac{1}{4}$ of 100, $\frac{1}{6}$ of 24, $\frac{1}{8}$ of 48 etc. to work out as quickly as possible.

C Other fractions of numbers (p 119)

T

Some teachers have found it helpful if pupils lay out their calculations as in the two examples at the top of page 119.

B Simple fractions of numbers (p 117)

B1 (a) 2   (b) 7   (c) 11
    (d) 5   (e) 18

B2 (a) 2   (b) 8   (c) 6
    (d) 10   (e) 13

B3 P and D   Q and C   R and E
    S and B   T and A

B4 The pupil's diagrams

B5 (a) 8   (b) 7   (c) 9
    (d) 2   (e) 9

B6 (a) 782   (b) 493   (c) 7869
    (d) 52 341   (e) 259   (f) 25 897
    (g) 129 670   (h) 456   (i) 633

B7 About 3496

B8 500 000

B9 8000

B10 12 500

C Other fractions of numbers (p 119)

C1 (a) 18   (b) 24   (c) 30
    (d) 3   (e) 33

C2 (a) 16   (b) 20   (c) 12
    (d) 14   (e) 22

C3 P and D   Q and F   R and A
    S and C   T and B   U and E

C4 (a) 27   (b) 12   (c) 15
    (d) 35   (e) 40

C5 16

C6 (a) 160 cm   (b) 24 years old

C7 (a) 528   (b) 3688
    (c) 2800   (d) 369 000

C8 128 m

C9 300 000

C10 (a) 68 000 km^2 (b) 340 000 km^2

C11 600 leaves

What progress have you made? (p 120)

1 (a) 9   (b) 5   (c) 6

2 (a) 9   (b) 5   (c) 10

3 (a) 55   (b) 200   (c) 126

4 (a) 14   (b) 60   (c) 21

5 (a) 128 688   (b) 2688
    (c) 513   (d) 840

Practice booklet

Section A (p 58)

1 They each get $\frac{1}{4}$ of a bar.

2 $\frac{2}{6}$ or $\frac{1}{3}$ of a bar each

3 $\frac{4}{6}$ or $\frac{2}{3}$ of a bar each

4 $\frac{1}{2}$ of a bar

5 (a) 2   (b) 4   (c) 10

6 (a) 3   (b) 6   (c) 12

7 (a) A person at table Q
    (b) They get $\frac{3}{4}$ of a bar each.

8 (a) The people at table Q – they get $\frac{3}{5}$ of a bar each.
      At table P they get $\frac{1}{3}$ of a bar each.
    (b) The people at table Q – they get a whole bar each.
      At table P they get $\frac{1}{2}$ a bar each.
    (c) The person at table Q – he or she gets 2 bars.
      At table P they get $1\frac{1}{2}$ bars each.

9 (a) You get most at table Q ($\frac{1}{2}$ a bar).
      At table P you only get $\frac{1}{3}$ of a bar.
    (b) You get most at table Q ($\frac{3}{4}$ of a bar).
      At table P you only get $\frac{2}{3}$ of a bar.

10 6 people

Section B (p 60)

1 P and B   Q and A   R and D   S and C

2 A $\frac{1}{3}$ of $12 = 4$ B $\frac{1}{5}$ of $15 = 3$

C $\frac{1}{7}$ of $14 = 2$ D $\frac{1}{3}$ of $18 = 6$

E $\frac{1}{2}$ of $12 = 6$

3 The pupil's drawings showing these amounts:

(a) 5   (b) 5   (c) 8   (d) 4

4 (a) 7   (b) 5   (c) 4   (d) 8

(e) 7   (f) 9   (g) 7   (h) 6

(i) 2   (j) 7   (k) 4   (l) 4

Section C (p 61)

1 P and A   Q and E   R and C   S and B
T and F   U and D

2 (a) 27   (b) 36   (c) 18   (d) 16

(e) 9   (f) 25   (g) 21   (h) 16

(i) 3   (j) 30   (k) 180   (l) 300

3 (a) 4   (b) 16   (c) 24   (d) 30

4 (a) 10 m   (b) 30 m

(c) 24 m   (d) 25 m

26 Negative numbers

Pupils order and find differences between negative numbers using time and temperature scales. They also interpret graphs and charts of temperatures.

p 121 **A** Lift-off!		Ordering events on a scale of time using negative numbers
p 122 **B** Freezing cold		Finding the differences between numbers using temperature scales
p 125 **C** Temperature graphs		Interpreting graphs involving temperatures

Optional
Sheets 187 and 188, scissors
OHP transparencies of sheets 187 and 188

Practice booklet pages 62 and 63

A **Lift-off!** (p 121)

The object of these class activities is to put negative numbers in order.

Optional:
Sheets 187 and 188, scissors, transparencies of sheets 187 and 188

'We wrote the times on paper and the pupils hung them on a "washing line".'

◊ You could begin the initial activity by asking some questions which familiarise pupils with the events, such as

How long before lift-off is the cabin hatch closed?

If the number line is drawn on an OHP or blackboard, pupils can be asked to come forward and mark the events on this. Alternatively sheet 187 can be used in either of two ways:

• Pupils cut out the events and place them in order.

• An arrow can be drawn from each event to where it lies on the number line, from which the order can be established.

◊ Pupils could start finding differences if you ask them how long it is between various events. For example,

How long after closing the hatch is the cabin checked for leaks?
(10 minutes)

How long after the decision to go is the Shuttle 40 km high?
(11 minutes)

Seconds away! (p 122)

This is another activity where pupils are asked to order events.
Sheet 188 can be used in the same way as sheet 187.
The correct order is G, O, C, P, R, F.

This example could be used for more questions on differences.

B Freezing cold (p 122)

Pupils have to read thermometers and calculate differences between temperatures.

◊ A maximum/minimum or freezer thermometer would be a useful aid here. A max/min thermometer has markers which show the highest and lowest temperature occurring since it was last set.

C Temperature graphs (p 125)

Pupils extract information from graphs.

◊ Information on maximum and minimum temperatures from a school weather station, regional weather centre, internet source or local enthusiast would make a useful stimulus for discussion.

B Freezing cold (p 122)

B1 (a) Glasgow 1°C   Paris ⁻10°C
London ⁻3°C   Oslo 0°C
Moscow ⁻5°C

    (b) Paris   (c) Glasgow

B2 ⁻4°C

B3 ⁻12°C

B4 3°C

B5 ⁻1°C

B6 ⁻4°C

B7 ⁻4°C

B8 ⁻8°C

B9 2°C

B10 (a) P 5°C, Q ⁻5°C, R ⁻25°C, S ⁻5°C

    (b) R

B11 (a) Freezer   (b) Fridge

    (c) Neither   (d) Fridge

B12 (a) Budapest   (b) 14 degrees

    (c) 17 degrees (d) ⁻8°C

B13 A temperature of **5**°C is **10** degrees higher than a temperature of ⁻**5**°C.

B14 A temperature of ⁻**5**°C is **10** degrees lower than a temperature of **5**°C.

B15 (a) 7 days   (b) 13 degrees

B16 (a) True   (b) False

    (c) True   (d) True

© **Temperature graphs** (p 125)

C1 (a) January and February

(b) June, July and August

(c) 9 degrees

C2 (a) 25°C   (b) ⁻5°C

(c) January and February

(d) 8 degrees   (e) 7 degrees

(f) October (12 degrees)

What progress have you made? (p 126)

1 ⁻32°C, ⁻14°C, ⁻3°C, ⁻1°C, 2°C, 15°C

2 (a) 23 degrees   (b) 5 degrees

(c) 8°C

3 (a) ⁻3°C   (b) 6 degrees

(c) Between 9 p.m. and 11 p.m.

(d) 2°C

Practice booklet

Sections A and B (p 62)

1 ⁻32 s, ⁻10 s, ⁻5 s, ⁻3 s, 2 s, 11 s, 31 s

2 5°C

3 ⁻10°C

4 6°C

5 ⁻9°C

6 ⁻36°C

7 A temperature of **5**°C is **20** degrees higher than a temperature of ⁻**15**°C.

8 44 degrees

9 (a) January and December

(b) July   (c) 20 degrees

Section C (p 63)

1 7°C

2 ⁻2°C

3 Wednesday

4 Saturday

5 (a) 5 degrees   (b) 7 degrees

(c) 7 degrees

6 Wednesday (9 degrees)

Review 4 (p 127)

1 (a) 7 cm and 5.4 cm

   (b) 3.1 cm and 4.4 cm

2 (a) $\frac{1}{3}$ red and $\frac{2}{3}$ yellow

   (b) $\frac{5}{9}$ red and $\frac{4}{9}$ yellow

   (c) $\frac{5}{10}$ or $\frac{1}{2}$ red and the same yellow

   (d) $\frac{2}{8}$ or $\frac{1}{4}$ red and $\frac{6}{8}$ or $\frac{3}{4}$ yellow

3 The pupil's enlargements

4 (a) 6   (b) 15   (c) 8   (d) 6

   (e) 7   (f) 5   (g) 10   (h) 6

   (i) 12   (j) 18

5 Harriet is 6 years old and 100 cm tall.

6 (a) 12   (b) 18   (c) 4   (d) 12

   (e) 34   (f) 10   (g) 12   (h) 16

7 (a) 8   (b) 5   (c) 2   (d) 4

   (e) 20   (f) 10

8

No of birds	Tally	Frequency
0–9	////	4
10–19	ﾞﾞﾞﾞ	6
20–29	ﾞﾞﾞﾞ	5
30–39	ﾞﾞﾞﾞ	5
40–49	ﾞﾞﾞﾞ	5
50–59	///	3
	Total	28

The pupil's bar chart

9

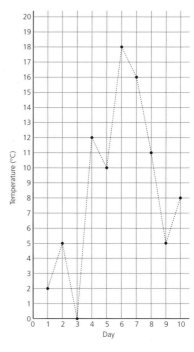

10 (a) 18°C   (b) 10°C   (c) 6°C

    (d) ⁻14°C   (e) ⁻5°C   (f) 0°C

11 (a) $5\frac{3}{4}$ hours or 5 hours 45 minutes

    (b) 13 degrees

    (c) $4\frac{3}{4}$ hours or 4 hours 45 minutes

    (d) 10 degrees

    (e) 7:05 p.m.

Mixed questions 4 (practice booklet p 64)

1 (a) 12   (b) 4   (c) 10   (d) 20

   (e) 6   (f) 12   (g) 8   (h) 16

2 (a) 16   (b) 12   (c) 10   (d) 4

   (e) 3   (f) 7   (g) 8   (h) 20

3 (a) 10   (b) 4   (c) 16

   (d) 10   (e) 5   (f) 5

4 (a) 12°C is **2** degrees higher than 10°C.

   (b) ⁻**2**°C is 8 degrees higher than ⁻10°C.

   (c) ⁻8°C is 4 degrees higher than ⁻**12**°C.

   (d) ⁻4°C is 4 degrees **lower** than 0°C.

5 (a) 1994   (b) 1997 and 1998

   (c) 1998 and 1999

   (d) 1999 and 2000